T0229874

VLSI RISC Architecture
and Organization

ELECTRICAL ENGINEERING AND ELECTRONICS

A Series of Reference Books and Textbooks

EXECUTIVE EDITORS

VLSI RISC Architecture and Organization

Stephen B. Furber

Acorn Computers Limited
Cambridge, England

Marcel Dekker, Inc.   •   New York and Basel

Library of Congress Cataloging-in-Publication Data

Furber, Stephen B.
    VLSI RISC architecture / Stephen B. Furber.
        p. cm. -- (Electrical engineering and electronics ; 56)
    Includes bibliographies and index.
    ISBN 0-8247-8151-1 (alk. paper)
    1. Reduced instruction set computers.   2. Computer architecture.
    3. Integrated circuits--Very large scale integration.   I. Title. II. Series.
    QA76.5.F865   1989
    004.2'2--dc19   89-1618
                                                        CIP

This book is printed on acid-free paper.

MARCEL DEKKER, INC.
270 Madison Avenue, New York, New York 10016

Current printing (last digit):
10 9 8 7 6 5 4 3 2 1

PRINTED IN THE UNITED STATES OF AMERICA

to Val

Preface

This book is written for final-year undergraduate and postgraduate students of computer science, and for engineers who study or design computer architectures. It describes the main architectural and organizational features of modern mini- and microcomputers, and explains the RISC (*Reduced Instruction Set Computer*) philosophy against the background of continuing architectural development.

The evolution of computer architecture is driven by many forces. These include not only rapid advances in the underpinning technologies used to implement logic, memory and peripheral devices, but also advances in the conceptual abstractions which are used to enable thought and discussion about how best to exploit those technologies. These abstractions include hardware architectures and organizations, and software systems and languages. The debate which arose amongst computer architects about the merits of the RISC approach cannot be understood in isolation from these other factors, both concrete and abstract.

Since the RISC approach to computer design defies precise definition, it is explained here mainly by way of example. The historical designs which embodied the principle and led to its declaration are presented in detail, and the subsequent commercial designs which have exploited those ideas are introduced. Although this information already exists in the form of published papers and manufacturers' data books, the styles of presentation and terminologies used vary greatly. This makes gaining a detailed understanding of a design a lengthy process. Here the presentation and terminology will be kept uniform, and attention will be drawn to the novel and interesting features of the machines described.

The intention is not to present comparisons which establish that a particular processor is superior to all others. Indeed, no specific performance comparisons are presented. Performance is very dependent on technology, and any comparison would be specific to one time and one point in technological development. Instead, emphasis will be placed on architectural and organizational features; architectural advances have more lasting merit than performance based on raw process technology alone.

The principle restriction on the scope of the book is to limit attention to VLSI implementations of RISC processors. There are several successful RISC architectures which have been implemented using standard logic families, but the trade-offs in such designs are different from the trade-offs made when the target is a CPU on a single piece of silicon (or gallium-arsenide). There is now sufficient variety amongst VLSI RISC architectures to offer ample material for a work such as this, and the restriction reduces the architectural space to one which may be explored thoroughly.

To offer insight into the design issues which arise in developing a RISC system, a detailed description is presented of the VLSI RISC chip set developed at Acorn Computers Limited in Cambridge, England. This system is the simplest of the many commercial offerings now available, and is therefore a good introduction to the design process. The discussion will cover the various options considered during the design process, the basis for the decisions taken, and the implementation details as far as is instructive. The development of even a simple system such as this is a major undertaking, and it is vitally important to adopt a structured approach to the design and validation of all system components if project timescales are to be controlled.

Research is continuing into VLSI computer architecture. The final chapter of the book is concerned with the directions being taken in current research, and offers pointers to likely future developments. The expected annual doubling of processor performance for the next few years will come only in part from advances in semiconductor processing; much of the performance increase will have to come from architectural improvements. It is interesting to observe ideas from mainframe and super-computer systems migrating onto VLSI processors. The blending of these with new features, which are in some cases practical only because of the use of VLSI, offers a rich vein of investigation in years to come.

Acknowledgements

I am grateful for permission from many organizations to use their copyright material, in original or adapted form:

- Figures 25, 27, 28 and 29 and table 2 contain material which is copyright ©Digital Equipment Corporation, 1980. All rights reserved. Reprinted by kind permission of Digital Equipment Corporation.

- Section 2.2 contains material which is copyright ©1985 by The Massachusetts Institute of Technology. Used with permission. All rights reserved.

- Chapter 4 and sections 3.2 and 5.3 contain material which is copyright ©Acorn Computers Limited, 1987 and 1988. All rights reserved.

- Section 3.4 contains material from Gerry Kane, MIPS R2000 RISC Architecture, ©1987, pp. 2-1, 2-3, 2-5, 3-13, 6-1, 6-15. Adapted by permission of Prentice-Hall, Inc., Englewood Cliffs, New Jersey.

- Figure 66 and section 3.6 contain material copyright ©Advanced Micro Devices, Inc., 1987 and 1988. Reprinted with permission of copyright owner. All rights reserved.

- Figure 73 and section 3.7 contain material which is copyright © 1986 Hewlett-Packard Company. Reproduced with permission.

- Section 3.8 contains material which is copyright ©Motorola Inc., 1988, reprinted courtesy Motorola Limited. All rights reserved.

- Figures 85 and 86 and section 3.10 contain material which is copyright ©1987, 1988 Intergraph Corporation. All rights reserved.

- Figures 91 to 95 and table 15 contain material which is copyright © INMOS Limited and is reproduced with permission of INMOS Limited.

- Other material is copyright as credited in the text.

I am grateful to my family for their forbearance during the preparation of this book; to Acorn Computers Ltd. for affording me the facilities to prepare it; to Bob Antell, Lee Smith, Mike Muller, Kechil Kirkham, Gordon Stevens, Jim Sutton and Andy Smith for reviewing draft copies; to Roger Wilson and Hermann Hauser for causing me to become involved in RISC processor design, and to the Acorn R&D team for their enthusiasm and commitment which make it all worthwhile.

Stephen B. Furber

The text and figures for this book were prepared using Twin and D respectively; both programs were written by Roger Wilson. The page layout was prepared using GCAL, a program written by Phil Hazel at Cambridge University Computer Laboratory. All these programs were run on an Acorn Archimedes RISC based personal workstation. An Apple LaserWriter performed the phototypesetting, using the PostScript page description language from Adobe Systems, Inc.

Contents

1
The Evolution of Computer Architecture

A *Reduced Instruction Set Computer* (RISC) is a member of an ill-defined class of computing machines. The common factor which associates members of the class is that they all have instruction sets which have been optimized more towards implementation efficiency than members of the alternative class of *Complex Instruction Set Computers* (CISCs), where the optimization is towards the minimization of the semantic gap between the instruction set and a one or more high-level languages.

Though the difference in emphasis may be clear to a microprocessor designer, the resulting differences between RISC and CISC processors are often hard to identify. The introduction of the RISC approach has, however, been accompanied by a dramatic rise in the power of commercial microprocessors. For this reason it should be given a significant place in the history of architectural development.

One way to understand the significance of the RISC approach is to see it against the background of architectural development in which it first arose. In this chapter we will look at the history of computer systems, and examine some of the machines which introduced important new features that have become standard requirements on all subsequent general-purpose computers. We will also look at the underpinning technologies which are used to build computers, and see how their high rate of development is constantly shifting the ground upon which architectural decisions are based. We conclude this chapter with a description of the VAX-11/780, a very sophisticated CISC machine, and introduce by way of contrast the basis of the RISC approach.

This chapter sets the context for the RISC movement. In the second chapter we will examine the earliest RISC architectures which established the principles and formed the basis for commercial exploitation. Subsequent chapters contain descriptions of commercial designs which have to a greater or lesser extent been influenced by the RISC philosophy.

1.1 BASIC COMPUTER ARCHITECTURE

A computer is a machine which performs a series of calculations on a set of numbers. The instructions for the series of calculations are called the program, the numbers are the data. A *general-purpose* computer is so-called because relatively few assumptions are made at the design stage about the nature of the programs which will be run on it.

The earliest general-purpose computers were mechanical, but had features which are still incorporated in their modern electronic descendants. The Babbage Analytical Engine (figure 1) was proposed in 1834, and had a memory for the data items, a central processing unit (CPU), a program using technology developed for the automatic control of weaving looms, and output to a printer or a card punch. The instruction set included branching conditionally on the sign of a number, and instructions with three operand addresses (two source and one destination). The design was never built; though technically possible, it was not well matched to the capabilities of the day.

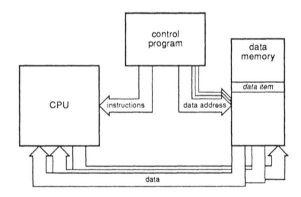

Figure 1: The Babbage Analytical Engine

1.1.1 The von Neumann Architecture

The earliest electronic machines capable of performing computations were laboriously programmed by changing the contents of a control store, in a manner analogous to changing the loom control cards in the Babbage design. The program was entirely separate from the data. A significant advance came with the removal of this separation, so that the program was kept in the same store as the data, and one program could be used to manipulate another. This *stored-program* concept is attributed to the designers of ENIAC, a vacuum tube based machine built for the US Army between 1943 and 1946. The concept was first expounded by von Neumann (1945), and incorporated into the IAS computer (at the Princeton Institute for Advanced Studies) which was completed in 1952. All general-purpose computers are now based on the key concepts of the von Neumann architecture (figure 2):

(1)   A single read-write memory contains all data and instructions.

(2)   The memory is addressable by location in a way which does not depend on the contents of the location.

(3)   Execution proceeds using instructions from consecutive locations unless an instruction modifies this sequentiality explicitly.

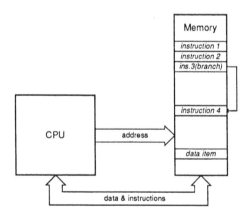

Figure 2: The von Neumann architecture

Though the von Neumann model is universal in general-purpose computing, it suffers from one obvious problem. All information (instructions and data) must flow back and forth between the processor and memory through a single channel, and this channel will have finite bandwidth. When this bandwidth is fully used the processor can go no faster. This performance limiting factor is called the *von Neumann bottleneck*.

1.1.2 The Harvard Architecture

A *Harvard architecture* (figure 3) has one memory for instructions and a second for data. The name comes from the Harvard Mark 1, an electromechanical computer which pre-dates the stored-program concept of von Neumann, as does the architecture in this form. It is still used for applications which run fixed programs, in areas such as such as digital signal processing, but not for general-purpose computing. The advantage is the increased bandwidth available due to having separate communication channels for instructions and data; the disadvantage is that the storage is allocated to code and data in a fixed ratio.

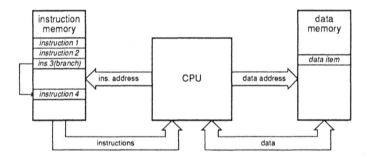

Figure 3: The Harvard architecture

Architectures with separate ports for instructions and data are often referred to by the term 'Harvard architecture', even though the ports connect to a common memory. For instance each port may be supplied from its own local cache memory (figure 4). The cache memories reduce the external bandwidth requirements sufficiently to allow them both to be connected to the same main memory, giving the bandwidth advantage of a Harvard architecture along with most of the flexibility of the simple von Neumann architecture. (The flexibility may be somewhat reduced because of cache consistency problems with self-modifying code.) Note that this type of Harvard architecture is still a von Neumann machine.

Although the Harvard architecture (modified or not) offers double the bandwidth of the simple von Neumann architecture, this will only allow double performance when the instruction and data traffic are equal. The

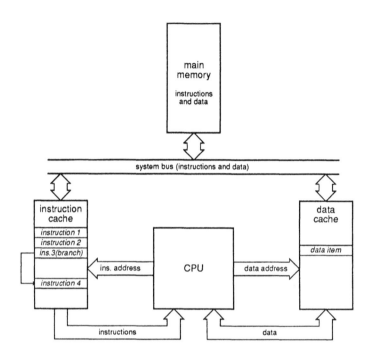

Figure 4: A modified Harvard architecture

VAX-11/780 has been used for a lot of measurements of memory traffic, and these measurements tend to suggest a reasonable match. RISC processors have two characteristics which make the match less good than the VAX. Firstly, they usually have less dense code, which increases the instruction traffic. Secondly, they are register to register rather than memory to memory architectures. This causes compiler writers to be much more careful about register usage, which can in turn result in much less data traffic. The use of register windows also eliminates considerable data traffic associated with procedure calls. A RISC CPU can typically require an instruction bandwidth six to ten times the data bandwidth (Patterson and Sequin, 1981), so that a Harvard architecture may only allow a ten percent speed-up (rather than the one hundred percent that would be suggested by the VAX statistics). Even so, the dual ported Harvard architecture has become a popular choice with RISC designers.

1.1.3 The Central Processing Unit

The Central Processing Unit (CPU) of a computer is the part that executes instructions and does the computing. It consists of one or more function units, some registers, and control logic. Typically in a VLSI processor there is one function unit which is an Arithmetic and Logic Unit (ALU), though there may also be a shifter. The registers will include the user visible registers, plus others which are not visible to the user, for instance the instruction register which holds the instruction which is currently executing. The control logic will be a mixture of PLAs, random logic, and possibly microcode ROM.

The part of the CPU through which data items flow is called the *datapath*. This includes the function units, and those registers which may contain data items. Datapaths are usually organized around *buses*, which are common routes for moving data from one register to another or to a functional unit. A 32-bit CPU will use 32-bit buses, so that all bits are moved simultaneously 'in parallel'. The number of available buses is an important determinant of cost and performance. The user registers are usually arranged into a regular block (or *bank*) where the design of the basic cell can be highly optimized for speed and space, and the size of the cell will depend on the number of buses which run through it. For simplicity and low cost, one bus would be ideal. From the performance point of view, however, a single bus would be a severe bottleneck. A typical datapath operation reads two registers, combines them in an ALU, and writes the result back to a register, requiring three accesses in all. Normally, therefore, the register bank will have at least two read buses and one write bus. (Some designs use the same physical bus for both reading and writing.) If a CPU is to achieve a data STORE operation in a single cycle using base plus index addressing (a normal addressing mode on a CISC processor), then it must read three registers in the cycle, which requires three read buses.

Datapath design therefore usually starts from the register bank. The number and width of the registers is generally fixed in the instruction set design (though not always; see the description of the Sun SPARC in chapter 3), as are the required operations and addressing modes. If the design calls for a very high number of register bits, then the pressure to keep the the number of physical buses low is great; if the number of register bits is moderate, it may be possible to gain performance by allowing many access buses. Once the register bus structure is decided, the function units must be connected to those or other buses, and an analysis performed of the data routes required by the various instructions. The bus structure is iterated until the cost/performance balance is appropriate.

A simple CPU datapath is shown in figure 5. The register bank has two read ports which are the sources of the ALU operands, and one write port for the ALU result. The ALU result may also be written to a *memory address register*. For a store instruction where the address has been set up in a register, that register is placed onto the B bus and fed to the memory address register via the ALU (which is configured for this operation to feed the B bus input directly to the result), and the register to be stored is placed onto the A bus and sent to the memory. A load could use the same route to generate the address, and then feed the loaded value from memory via the A bus and ALU into the destination register.

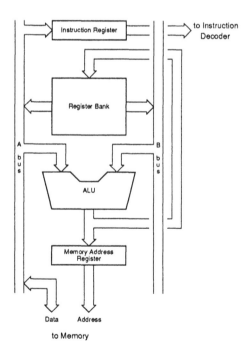

Figure 5: A simple CPU datapath

This datapath will handle internal operations and memory data accesses satisfactorily, but will be very inefficient at fetching instructions. Unless a separate instruction fetch unit is used, every new instruction will require an ALU operation to increment the program counter (which must be kept somewhere in the general register bank). If the instruction set is very complex and uses many cycles per instruction, the overhead of processing the program counter in the main ALU may be acceptable, but a RISC CPU is certain to contain some additional function unit (for instance an address incrementer) to reduce contention between the instruction fetch and execution activities.

Datapath design begins with a picture similar to this and then additional buses, registers and function units are added to avoid the bottlenecks which are identified in the course of determining the data routes for each instruction in the instruction set. It is useful to have a clear understanding of which instructions are most important to the eventual performance of the CPU, so that they are optimized in preference to less critical instructions. It is also useful to identify the theoretical minimum number of cycles for each instruction; if every memory transfer takes at least one cycle, an instruction which saves four registers will take at least four cycles, and this will usually allow considerable housekeeping activity to go on in the background at no additional cost in cycles.

Once the datapath is designed and the data routes have been determined for every instruction, control logic is added to decode instructions and to cause the datapath to perform the required actions. This will be illustrated for the case of the Acorn RISC Machine in chapter 4.

1.1.4 The Memory System

The memory system contains the user and operating system programs and data. It simply holds this information until the CPU wants it. If the CPU puts a value in a particular memory location, it will expect to find the same value there any time it comes back to look, unless it has overwritten it in the meantime or asked the I/O system to load something else into the same location. The memory is therefore an inert waiting area; all the action takes place inside the CPU.

The CPU communicates with the memory by means of:

- An *address bus*, which is driven by the CPU to tell the memory which location is required.

- A *data bus*, which may be driven by the memory (for *read* cycles) or by the CPU (for *write* cycles) to take the addressed data to or from the CPU respectively.

- *Control signals*, which specify whether the cycle should be a read or a write transfer, when it should start, when the data is valid, and so on.

The important parameters of a memory system include its size and its access speed. The size is largely a cost issue; pay twice as much, and you should be able to get twice as much memory. The speed is usually more fundamentally linked with the implementation.

Memory access speed has two components:

(1) *Latency.* This is how long it takes the memory to return the first word in response to a specific request for data.

(2) *Bandwidth.* This is how fast information can be supplied once the flow has started. It is the product of the rate of data transfer and the amount of data in each transfer, and is usually expressed in millions of bytes per second (Mbytes/s).

If program execution rate is determined by how fast instructions can be fetched from memory (as it often is), then the performance of sequential code will be bandwidth limited. The performance of a program composed entirely of branches will be limited by the memory latency. In the former case the memory can supply instructions in a steady flow, whereas in the latter case no sooner will a flow be established than a new request will be issued for a different flow. A real program will be somewhere between these two extremes.

Since memory access speed is such an important system parameter, it is worth looking at ways of optimizing it:

(1) *Very fast memory devices.*

The obvious way to improve both latency and bandwidth is to use faster memory devices. Unfortunately this has a direct impact on cost; the faster parts cost more. For a supercomputer this cost may be acceptable, but other classes of machine are constrained to use economical bulk memory technology. In most cases, this currently means *Dynamic Random Access Memory* (DRAM) devices. These parts typically allow 3 to 6 million random accesses per second, with a moderate price premium for the faster parts. Above 6 million accesses per second the technology changes to *Static Random Access Memory* (SRAM), and the price per bit jumps up significantly.

(2) *Wide data transfer buses.*

A wider bus will improve bandwidth (but not latency), because more information is delivered in the same time. Mainframe computers often use very wide buses for refilling caches. VLSI processor designers must take note of the significant additional cost of going from 32 to 64 data pins. The other disadvantage of a wide bus is that it is only effective when all the bits are required; if a single byte was required from a particular location then although a 64-bit bus offers more bandwidth, it does not improve the CPU performance (for that operation). However, accesses to isolated bytes are statistically infrequent, and usually a program will use several words from any area of memory which it accesses. This property of program *locality* would suggest that a 64-bit bus would almost double the useable bandwidth for most purposes compared with a 32-bit bus.

(3) *Fast memory modes.*

Although DRAM technology only supports up to 6 million random accesses per second, computer memory accesses are not truly random (as mentioned above), and DRAMs have faster localized access modes which may be exploited to improve memory performance.

DRAMs are implemented in a way that exposes their internal structure, which is typically a near-square array of bits (figure 6). A particular bit is specified by its row and column in the array, and standard devices accept the row and column addresses in multiplexed form through the same pins. Once a particular row has been specified, it is possible to access many columns in that row, and the access time is significantly reduced so long as the same row continues to be used. Therefore if the CPU requests four consecutive words from an address these can be delivered at significantly higher bandwidth than if it requested each individually. In general the usable memory bandwidth can be doubled, and it may be increased by a factor of six in some special cases. (Note, however, that memory latency is again not improved by this approach.)

(4) *Interleaved memory banks.*

If the memory is divided into two or more interleaved banks, then the available bandwidth may be increased and the latency reduced. Following one access, DRAMs have a recovery period before a subsequent access can start. If one bank is recovering while another performs an access, this recovery period can be hidden. The higher the number of banks, the lower the probability of consecutive random

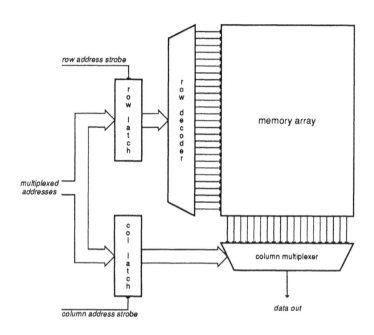

Figure 6: Dynamic memory read organization

accesses using the same bank and hence suffering a delay whilst waiting for the memory to recover. Because a high proportion of memory accesses are sequential, two interleaved banks will reduce the average latency by almost the memory recovery time.

If several consecutive words are required, all banks can begin accesses as soon as they have recovered from previous activity, and they can then send their respective data in rapid succession to the CPU. The achievable bandwidth is the bandwidth of one bank multiplied by the number of banks.

The disadvantage of interleaved memory is the cost and complexity of separate control circuitry for each memory bank. It is extensively used in mainframe computers, but has not been used much with VLSI CPUs, probably because memory access speed has only recently become the principle performance limiting factor of microprocessors. Cache memories are the most cost effective way of overcoming the memory bandwidth limitation initially, but as performance continues to rise the cache miss penalty becomes increasingly important, and memory interleave is one of the few options left once cache memories have been fully exploited.

Examples of the main memory characteristics of various machines are given in table 1.

Table 1. Memory characteristics of various computers

Machine, Class, Year	Memory Size (Mbytes)	Memory Width (bytes)	Memory Latency (ns)	Memory Bandwidth (Mbyte/s)
CRAY-1, Supercomputer, 1975	2-8	8	50	640
Archimedes, RISC PC, 1987	.5-4	4	250	25
VAX-11/780, Supermini, 1978	.25-12	4	1600	13.3
6800, Microprocessor, 1974	.064	1	1000	1

1.1.5 The Input/Output System

Computers use their CPUs to process data in memory according to instructions in memory. They must load the program and data into memory from somewhere before they can begin execution, and they must save the results somewhere at the end of the computation for subsequent use. The Input/Output system contains an assortment of devices which are the sources and/or destinations of this information.

The CPU typically sees the I/O world as a set of addressable locations, similar to memory. There may be special instructions for controlling I/O, or the standard memory referencing instructions may be used, but the idea is the same. The difference between memory and I/O is that while the memory simply stores what it is given until the CPU asks for it again, reading or writing an I/O register can have all sorts of effects on associated electrical and mechanical components, and there is no guarantee that if you read the location again it will return the same value. The programs which deal with the I/O hardware are *device drivers*, and these are often the most idiosyncratic sections of operating systems. These days they are frequently the only program segments still written in assembly code.

The I/O system may contain a variety of peripheral interfaces, each of which has its own characteristics. Some of the possibilities are:

- *Disk storage*. Disks have a high data transfer rate (5 or 10 Mbit/s for typical low-cost hard disks), and are heavily used, particularly in paging virtual memory systems. They have considerable rotational latency, but once the transfer commences the full bandwidth must be absorbed; the transfer cannot be stalled, except by accepting the full rotational delay again.

- *A local area network*. LANs also have a high data transfer rate (Ethernet uses 10 Mbit/s, and 100Mbit/s networks are becoming available). Unlike disks, they have the characteristic that they can initiate activity. A disk will only produce data when a transfer is initiated by the CPU, so the CPU can prepare for the reception. A LAN transfer may be caused by a request from another CPU on the network, so no specific preparation is possible. It is usually possible to stall a transfer - the protocol will retry if the first attempt fails. In a fully distributed system the LAN may be very active with other users running processes on your system. Otherwise the only frequent activity is to check every packet to confirm that it is not for you, and this may require CPU activity or may be performed in hardware.

- *A video display*. Video displays require extremely high bandwidth, for example a 1k by 1k pixel 256 colour display with a 60 Hz refresh rate requires an average 60 Mbytes/s of video data, and around 120 Mbyte/s peak. This is so high that a conventional memory system cannot support it, and special memory devices are required. This sort of video data will not be allowed to flow down the main bus, and the CPU will certainly not be involved in handling the data stream.

   Lower performance video displays may use part of main memory as a frame buffer, and access it via the processor bus. Hardware will be required to handle the data stream.

- *Serial lines, printers, etc.* These devices are very slow compared with CPU speeds, and are often controlled directly by software, including an interrupt task to supply the data stream.

These I/O functions are implemented in a number of different ways, according to cost/performance trade-offs. The video system design must be an early consideration in any computer system which is to generate a display directly, and the solutions are many, varied, and specialized. Other I/O may be supported by involving the CPU in control and data transfer activities (for which the hard disk and LAN systems may need sector and packet buffers), though this is likely to absorb a high proportion of the CPU power for the high bandwidth functions. Alternatively, the CPU may set up control registers but leave the data transfer to a *direct memory access* (DMA) controller. The DMA controller will move the data directly from the peripheral controller to the memory (or vice-versa), absorbing bus bandwidth but otherwise leaving the CPU free for normal processing. At the end of a transfer an interrupt will inform the CPU of the completion, whereupon the CPU can re-initialize the DMA system to perform a subsequent transfer. The third level of I/O system uses a channel controller, which is a simple processor capable of independently executing a sequence of I/O commands from memory. The CPU only has to generate an appropriate command sequence, and the channel controller will do the rest.

All of these systems must generate interrupts to request CPU attention, and the simplest system may generate a very large number of them. If a CPU is intended for use in such a simple I/O environment, it must have a low interrupt overhead. If, for instance, it automatically saves a large number of registers on interrupt entry, this is likely to take several microseconds, and limit the maximum interrupt frequency. A CPU targetted for good interrupt response is likely to have registers reserved for interrupt routines, so that little or no register saving is required.

An important factor in interrupt response is the interrupt latency, which is the worst case time from an interrupt being requested to the interrupt handling routine being entered. Multi-cycle instructions make this worse, unless they are interruptable in mid-instruction, since the worst case calculation must assume that the longest instruction has just been committed when the interrupt request arrives. A complex entry sequence is also bad. RISC CPUs should have a head start in interrupt performance because they don't in general have long multi-cycle instructions or complex entry sequences, but this area of the CPU requires very careful design.

1.2 ADVANCING TECHNOLOGIES

Computers are systems built from electronic components. The techniques used to make electronic components have improved rapidly over the short period of computer history, and these improvements have repeatedly shifted the ground underlying the assumptions upon which architectural decisions are made. There are many ways of characterizing the improvements, but most relate to some aspect of integrated circuit technology, as integrated circuits have become the universal building blocks from which computers are made.

Figure 7 shows how the characteristic size of a feature (for instance a transistor, or the spacing between two metal tracks) has reduced over the years, and is expected to continue to do so for the next few years at least. This figure shows the trend for memory devices; logic processes are usually a year or two behind. As the feature size reduces, it is generally possible to get more circuitry (bits of memory, or logic gates) on one integrated circuit. Other factors may make it impossible, the most common of which will be the heat generated by the extra circuitry. This has caused a general move towards logic families where gates which are not switching do not generate heat, and as a result the amount of circuitry has been able to rise in line with the reduced feature size.

In addition to allowing increased circuitry on a chip, a smaller feature size tends to increase the speed of the circuitry. This is a result of the fact that on-chip switching speeds are determined mainly by the capacitance of the interconnect from one stage of circuitry to the next (for long connections such as buses) or the capacitance of the gates in the next stage (for short connections between neighbouring gates). Both these capacitances are reduced by shrinking the process. The reduced feature size has benefits on all sides, and the continuing improvement in this area is an important enabling factor in the drive for higher performance computers.

A computer may be thought of as logic (which processes information to produce results), memory (where the results are stored until they are required again), and peripherals (disks, printers, and the like). The different characteristics of these system functions result in different technological dependencies.

1.2.1 Logic Technologies

The basic unit of computer logic circuitry is the Boolean gate, which performs a simple Boolean algebraic operation (AND, OR, NAND, NOR, etc) on a number of inputs to produce an output. In fact, any logic function

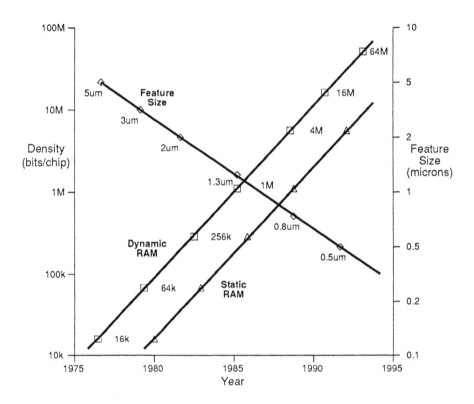

Figure 7: Memory density and feature size vs. time
(From Hitachi, 1987)

may be generated using only 2-input NAND (or NOR) gates, though simpler and faster circuits are possible if a wider choice of gates is available. It is also possible to build register latches using standard gates (to separate consecutive stages of a sequential logic structure), though more efficient latches can usually be built with transistor structures not based on gates.

The earliest integrated logic families were based on bipolar transistors, and the TTL (*Transistor-Transistor Logic*) 7400 series introduced by Texas Instruments rapidly established a widely used standard. This standard includes a 5 volt power supply, and high and low logic levels which are optimized for this asymmetric bipolar technology. The success of this logic family can be gauged from the fact that entire computers have been built from CMOS (*Complementary Metal Oxide Semiconductor*) chips which operate from 5 volt supplies and use TTL logic levels, despite the fact that these are very unsuitable for modern CMOS technology.

ECL (*Emitter-Coupled Logic*) provides higher speeds at higher power, but even TTL uses too much power per gate (whether switching or not) for VLSI applications. The first viable VLSI technology was PMOS (*P-type MOS*), which is based on FETs (*Field Effect Transistors*). This was rapidly overtaken by the higher performance NMOS (*N-type MOS*). NMOS gates still dissipate a little static power, though only in one of the two possible output states. As VLSI circuits incorporate more gates on a chip, even this static power becomes a problem, and the disadvantage of CMOS (extra process complexity) have become outweighed by its advantage of zero static power dissipation.

A CMOS 2-input NAND gate is built out of two n-type and two p-type FETs, as shown in figure 8. When an input is low, it turns on its p-type transistor which pulls the output high. It also turns off its n-type, and since FETs have an extremely high 'off' resistance, no DC current flows. The only power used is that required to charge up the output node. Conversely when both inputs are high, both p-types are off, but the n-type stack is on. The output is therefore pulled low, returning its stored charge to ground through the n-type stack.

The standard CMOS gate is fine for most applications, especially when it does not switch very often, because when not switching it consumes no power. It does, however, have a drawback in speed critical applications, which is that each logic gate loads each input with two transistor gates. The transistor gate loads represent a high proportion of the capacitive load on the previous gate, so if only one transistor were to be driven the previous gate would operate significantly faster. Various schemes have been devised for achieving this, mainly based on dynamic logic structures. In figure 9 another CMOS 2-input NAND gate is shown. Here a clock is used to precharge the output, during which time the inputs are assumed to be low. Then the charging transistor is turned off. If both inputs subsequently go high at the same time the output will be discharged, otherwise it will remain high.

A complex logic structure may be built up of cascades of such gates, but it must be designed in such a way that input transitions only occur in the right time period and in the right sense. The inputs of this gate must only make positive transitions, whereas the output can only make a negative transition. Therefore either an inverter must be inserted between stages, or the next

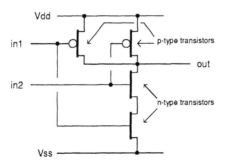

Figure 8: A CMOS static 2-input NAND gate

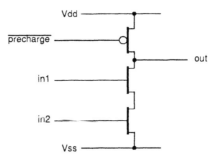

Figure 9: A CMOS dynamic 2-input NAND gate

stage must be built as a complementary structure which requires and produces the opposite transitions. Figure 10 illustrates how two 2-input NAND gates may be cascaded, and here the precharge control is also used to ensure that the first stage rises to Vdd independent of the logic levels on its inputs. The gates will work correctly so long as in1 and in2 are stable or only make transitions from Vss to Vdd during the active period. It is assumed that in3 is generated by a similar first stage gate.

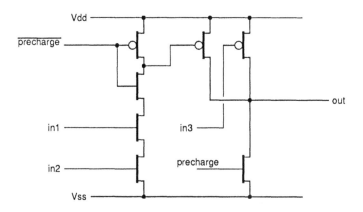

Figure 10: Cascaded CMOS dynamic NAND gates

This alternating complementary gate structure is called *zipper logic*, and is widely used for constructing time critical logic functions such as the ALU in a microprocessor. It is faster and uses fewer transistors than an equivalent static logic implementation, but requires very careful control of the timing of input transitions.

The downward trend in feature size allows the amount of logic on a single chip to increase, and the propagation delay of a typical gate to go down. At present (1988) the most advanced logic processes in mass production use feature sizes around 1 micron, on which typical gate speeds are below 1ns, and gate counts are up to 100,000 on a chip. The logic processes are usually a year or two behind memory processes in terms of feature size, but follow the same development line. The number of gates will scale approximately with the number of memory bits, and the propagation delay of a gate is roughly proportional to the feature size. Therefore in 1995 we may expect logic chips with 3 million gates which switch in 200ps built on 0.5 micron processes. These will enable very much more powerful CPUs to be built than are possible using present technology.

1.2.2 Memory Technologies

We have already looked at the trends in memory technology, which show the number of bits on a chip doubling every 18 months. The memory manufacturers can see how to maintain this remarkable rate of progress at least until the mid 1990s, by which time the amount of memory on one chip will exceed the total memory of a typical 1970s mainframe computer.

There are two basic types of semiconductor read/write memory. *Dynamic* memories store the information as electrical charge on a capacitor, whereas *static* memories store the information in a bistable flip-flop. The dynamic memory cell requires one transistor and a capacitor for each bit of data, and is therefore denser than the static memory cell which requires four transistors plus either two resistors or two transistors of a different type for each bit.

The dynamic memory cell (figure 11(a)) connects the bit line to the storage capacitor through a pass transistor which is controlled by the word line. During a write operation, the bit line is driven high or low and the capacitor follows suit. The word line is then deactivated, turning off the pass transistor and isolating the charge on the capacitor. During a read operation the word line is activated, but the bit line is not driven. Instead, the charge on the capacitor (if any) is bled onto the bit line and detected by a sense amplifier. The read operation is destructive, and the cell must be rewritten for future use. The charge on the capacitor will leak away over a few milliseconds, so every cell must be read and rewritten (*refreshed*) periodically if the data is to be retained for any length of time.

The static memory cell (figure 11(b)) uses a cross-coupled pair of inverters for data storage. During a write operation, the true and inverse bit lines are driven to appropriate levels, and the word line switches both pass transistors on. The low bit line will overpower the inverter pull-up within the cell to force it to the correct state. When the word line is deactivated the positive feedback within the cell ensures that the data is retained permanently. When the word line is activated for reading, the inverters drive the bit lines through the pass transistors. A differential amplifier on the bit lines can detect the stored value very quickly. Read out is not destructive, and no refresh activity is required.

The static cell as drawn requires both n-type and p-type transistors, and in normal CMOS processing these different types must be in isolated areas (wells) of the substrate, which causes the cell size to be quite large. Many manufacturers use a special process for SRAM which includes a very high resistance polysilicon which is used to form pull-up resistors to take the place

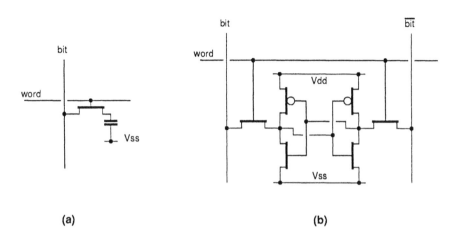

Figure 11: Basic (a) dynamic and (b) static memory cells

of the p-type transistors. All the transistors in the cell are now n-type, and can occupy the same well. Future CPUs with large on-chip caches may require this additional process to minimize the area taken by the cache.

DRAM is denser than SRAM, and has a much lower cost per bit, but it is also significantly slower. DRAM is therefore the usual choice for main memory, and SRAM for cache memory. The trend in DRAM is mainly to higher capacity; the access time is improving only marginally, though local access modes are getting faster (with successive accesses restricted to the same row of memory bits). Video DRAMs (VRAMs) are extreme examples of fast local access, as they allow very high bandwidth serial access for supporting raster displays. SRAMs are following both higher capacity and higher speed development paths.

1.2.3 The Impact of VLSI

There are several particular considerations which must be born in mind when designing a computer architecture for VLSI.

(1) *Chip size.* The cost of manufacture of an integrated circuit is closely related to the area of the chip. Integrated circuits are formed on thin circular wafers of silicon cut from single crystals. The wafer is several inches in diameter, and several circuits (often hundreds) are formed on it at the same time. The wafer is cut into the individual die after

fabrication and testing. The cost of processing a particular size of wafer is roughly constant, but defects cause die costs to increase more rapidly than linearly with area. Chips above a certain area are effectively unmanufacturable. Therefore the functionality which is to go on a single chip must be specified very carefully to ensure that the economics of the product are appropriate. The designer must also be aware of the strongly two-dimensional nature of VLSI design, which makes the topology of the functions and connections between them extremely important. Topologically efficient layouts (such as most memory devices) can contain an order of magnitude more transistors per unit area than interconnect dominated random logic structures. Estimating the eventual size of a layout from a logic diagram is hard.

(2)   *Pins.* The number of I/O pins is an important determinant of chip cost. Every bond wire (which connects a pin on the package to a bond pad on the chip) is a potential source of manufacturing failure, and packages with large numbers of pins tend to be expensive. On the other hand, bandwidth is a determinant of performance, which is also economically important, and high bandwidth requires a lot of pins. A lot of fast switching outputs require several power and ground pins to limit transients in the power and ground rails on the chip, and these must be allowed for in the pin budget.

(3)   *Power.* Chips must operate within specified temperature ranges, and large fast switching chips generate a lot of heat, even when they are built from CMOS. This heat must be removed, and the removal has a cost associated with it. Up to a watt of heat can usually be handled by a cheap (plastic) package in a convected airflow; much over a watt and a more expensive (ceramic) package is required, and forced airflow (or even water-cooling) may be needed. If excessive heat affects the design of the equipment casing or restricts the application environment then power can be a major design issue.

(4)   *Regularity.* An important aspect of VLSI is that the effort required to produce a design does not depend on the total number of transistors used so much as the number of different elementary (*leaf*) cells required. Once a single register bit has been designed, it is relatively easy to replicate it to produce a register bank of arbitrary size. It is certainly quicker than generating a random logic function which uses far fewer transistors.

Therefore VLSI designers strive to build *regular* structures, which are based on tessellated arrays of leaf cells. Datapaths are good because each bit will be identical in most respects to the others, but carry look-ahead circuitry is bad because it tends to break this regularity (repeating typically only every four bits across the ALU). Random logic is always a problem, but can often be implemented in ROM or PLA (programmable logic array) structures, where the overall structure is regular and the function is defined by the presence or absence of a minor feature in the array. These structures are usually generated automatically from a tabular or Boolean expression of the desired function, and are therefore very easy both to layout and to modify if necessary. The only drawback to these regular implementations of random logic is that they tend to be slower than an implementation based on gates.

The intermediate approach to random logic is to use a cell library, where a set of standard gates, latches, flip-flops, etc. is designed and characterized, and the circuit designer picks up appropriate cells and wires them together. This approach does not yield the smallest possible layout, but the speed can be good, and it may be a good way to implement the small amount of logic which for one reason or another cannot be incorporated into a regular structure.

(5) *Correctness.* It is hard to work out what is wrong with a VLSI device if it does not work, and trying out a fix takes several months. There is therefore a great incentive to produce working devices first time. When the device has the complexity of a 32 bit processor, this is not easy.

(6) *Testability.* Once the design is correct, it is tempting to assume that mass production is straightforward. This assumption is false! When a large integrated circuit is manufactured in volume, at least half of the manufactured devices will not work. The defects which cause failure are varied and random, and all devices must be thoroughly tested to identify the rejects. The test program should exercise all the transistors on the chip to ensure they are functioning, and make sure that any failure will affect the measured outputs at the pins. Speed critical paths must be measured to make sure that the transistors are up to the required strength, and so on.

The design and development of the test program absorbs as much effort as the logic design of the device itself, and can be greatly eased by careful consideration of test issues during the logic design phase. Considerable quantities of logic may be added just to simplify testing. Testing costs can be a significant proportion of the total device costs. Designing for testability is vital.

These considerations affect architectural decisions in complex ways, and furthermore they are tied to semiconductor technology which is advancing rapidly. The constraints represent moving targets - a VLSI architecture which is in some sense optimal at one point in time will cease to be so a few years later, and would have been unmanufacturable a few years earlier.

It is necessary, therefore, when designing an architecture for VLSI, to estimate the time to complete the design and to implement it, and to match the design to the semiconductor technology which might reasonably be expected to be available at that future time. A short design and implementation cycle makes such prediction more reliable, and this is one of the advantages claimed for the RISC approach to processor design.

1.3 ARCHITECTURAL ADVANCES

Over the short history of computer development, many significant steps forward have come not from advances in base technology, but from changes in the use of existing technology. Some of the key advances are described below.

1.3.1 Powerful Instruction Sets & Microcode

Microcode, or microprogramming, was first proposed by Wilkes and Stringer (1953) as a way of making the logic which is used to control the function of the CPU more regular. It enables a simple datapath to perform a complex instruction, and it achieves this by breaking the complex function down into a series of simple operations which are performed sequentially.

The basic microcode structure is shown in figure 12. A microcode address is formed from some or all of the contents of the instruction register, together with some state values which are internal to the micro-control unit. This address is decoded to drive a unique row of a matrix, the columns of which are the control signals for the datapath. The value of each control for a given microcode address is determined by the presence or absence of a diode at the point in the matrix where the row and column cross.

A second matrix generates the internal state values which are used along with the instruction to determine the microcode address for the next cycle. If the required sequence depends on the outcome of a test, the condition bit which contains the result of the test may be used to select one of two alternative rows in the second matrix.

The advantage of microcode is that the control logic is implemented in a very regular structure, which these days is usually a ROM. Changes or additions to the instruction set require only that bits in the ROM are altered, or possibly some more ROM has to be added. For even greater flexibility all or part of the ROM could be replaced by writeable RAM.

Some microcoded machines allow microprograms to use branches and subroutines, and the microsequencing logic can become very complex. Some of the techniques used in microprogramming have been adopted by designers of some RISC processors, but this time at the user program level. These borrowed techniques include delayed branches and software managed pipeline interlocks.

The IBM System/360 range illustrates some of the benefits of microcode. Here an instruction set was designed for high functionality on the top-end

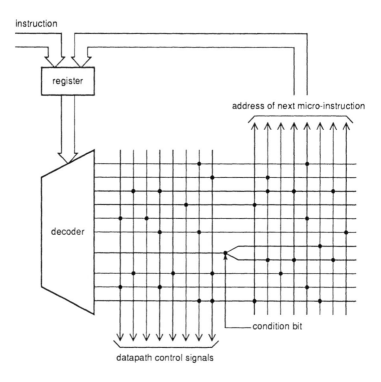

instruction

register

address of next micro-instruction

decoder

condition bit

datapath control signals

Figure 12: A microcoded control unit
(From Wilkes and Stringer, 1953)

machines. The top-end machine was not microcoded, as it made extensive use of multiple concurrent functional units, and the interdependencies between these precluded the use of microcode. The power of microprogramming is demonstrated lower down the range where microcode was used to enable much simpler processing units to retain instruction compatibility with the high end machine (Stevens, 1964), thereby allowing a compatible series of machines to be built covering a very wide range of cost/performance points.

A feature of microcode is that the microcode ROM usually has very good access time, so frequently used operations will run fast if they are microcoded. One of the earliest RISC exponents (Radin, 1983) pointed out that cache memories also have good access times, and whereas microcode only contains the static set of operations chosen by the original designer, a cache can contain a dynamic set of frequently used operations selected automatically by the hardware to suit the current task.

The RISC movement has been seen as a reaction against microcode. This may be misleading. It is the complex instruction sets which are called into question, not the microcode implementation. Although it is true that in small machines microcode is needed to support complex instruction sets, it is equally possible to microcode a simple instruction set. Early RISC designs avoided the use of microcode, but some later commercial RISC designs have reintroduced it. The RISC trend to single cycle execution of most instructions will preclude the use of the more exotic microcode constructs. With a single cycle instruction, the microcode ROM is really just a regular decode structure.

1.3.2 Virtual Memory

Applications expand to fill the space available to them. The quest for ever larger memories led to an architectural advance in 1961, when the prototype Atlas computer was commissioned at Manchester University, England (see Kilburn et al, 1962). The *single level store* concept embodied in this machine was the forerunner of the ubiquitous *virtual memory* systems used today.

The Atlas had 16,000 words of core store which was the processor's main memory. The programmer, however, could work under the illusion that he had 100,000 words of main memory. Most of this 100,000 words was resident on a magnetic drum store, and was only copied into the core store when it was required by the program. The operating system had sophisticated algorithms for determining which parts to keep in core and which to leave or copy out to disk.

The user's address space is divided up into pages of 512 words, as is the core store. Each page in the core store may be loaded with any of the user's pages at any time. The correspondence is maintained in a *content addressable memory* (CAM), which is a set of 32 page address registers with comparators. The page address register for core page N will contain the number of the user page which currently resides in core page N. When the user page is presented to the CAM, it will match page address register N, and the CAM will output N and indicate a *hit* . If a user address is presented which does not match any page address register, *hit* will not be indicated. This will cause a trap to the operating system, which will load the required page into core, update the page address register, and resume the user

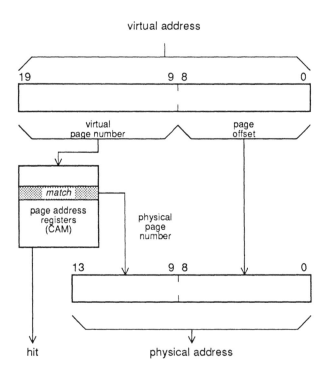

Figure 13: The Atlas single level store

program. This operating system activity is completely transparent to the user program, apart from a delay while the page transfer takes place.

There are now many variations on the virtual memory theme, including the use of segmented address spaces where code resides in one segment and data in another, and each segment has its own page table and protection level. The translation tables for modern systems can be very large, and are kept in main memory. Page tables are often built as two levels to reduce the amount of memory they occupy, and they can be kept in virtual memory and themselves paged out onto disk. The issues and options are largely the same for both RISC and CISC processors. We shall look at the VAX-11/780 translation scheme at the end of this chapter, and consider the virtual memory support on the various RISC CPUs in subsequent chapters. To avoid every memory reference requiring an additional access for the address translation, most systems include a *Translation Look-aside Buffer* (TLB), which is a cache of recently used page table entries. Quite a small TLB (64 or 128 entries) will have a good enough hit ratio for most purposes.

The idea of separating the storage into two hierarchical levels which look the same to the user has been applied widely, and will recur when we look at instruction and data caches. The idea of making the physical machine constraints (in this case the physical memory size) invisible to the user is also very powerful. As the program exceeds the constraint the degradation in performance is less sudden. With a physically addressed machine, a program which equals the available memory size runs perfectly, but add one instruction and it can't be run at all. In a machine with virtual memory, a program equal in size to the available memory will run somewhat slower than in the physically addressed machine, but the additional instruction will make very little difference.

Virtual memory is widely used, but it is not a panacea. A very large program will not run satisfactorily on a very small machine, however clever the paging algorithms are. Every program has a characteristic *working set*, which is made up of those parts of the program which are used during some period of time related to the disk transfer time. As the size of the working set of the program approaches the size of the available physical memory, the rate at which pages are swapped will increase rapidly and the program will run unacceptably slowly. A typical program may use a total address space up to twice the physical memory size and still run satisfactorily.

Virtual memory requires the concept of address translation from virtual to physical addresses. Address translation is powerful for other reasons also, as we see in the next section.

The relevance of virtual memory to the RISC movement is that it has become almost mandatory to support it, at least in designs destined for commercial exploitation. The only commercial exception is the Inmos transputer, and we shall look into the Inmos approach in chapter 3. Supporting virtual memory can be complex; any instruction which may cause a page fault must be restartable, or special registers must be allocated to contain sufficient information for the page fault handler to complete the faulted transfer. Somehow the illusion that the memory was really there all the time must be maintained.

1.3.3 Multi-tasking

Until recently a computer has been an expensive piece of equipment, and the pressure to make best use of its capabilities has been intense. Running one program at a time is inefficient, as that program may only require a small fraction of the available memory, and the processor will be idle for long periods while it waits for information from a disk. If there were two programs resident, the CPU could work on one until a disk transfer was required, then switch to the other while the transfer took place, and then switch back to the first. The CPU is thereby better utilized. In practice several programs are kept active at any time, and the processor switches between them according to the availability of data, priority of task, etc (figure 14).

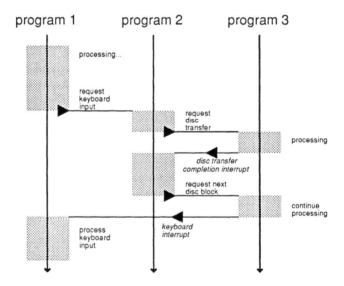

Figure 14: A typical multi-tasking thread of execution

With the advent of the personal computer and workstation the emphasis has switched. Rather than having several people working to keep one machine busy, it is now the machine which waits until required. It must respond as quickly as possible when called to do something, so as not to keep its user idle. PCs have been predominantly single tasking in the past. Now even here multi-tasking systems are gaining popularity and seem likely to become standard. The reason is that the PC is becoming involved in more aspects of its operator's business, and he/she would like to have multiple documents and spread-sheets visible at the same time, while the machine handles electronic mail in the background and continues to print his/her latest report. A multi-tasking operating system provides the best framework for the construction of this level of functionality.

Multi-tasking requires two things. Firstly, if it is to be at all flexible, it must allow a task to be loaded in different parts of store at different times. A machine with a fixed set of tasks loaded at fixed addresses is very inflexible. Secondly, it must allow tasks to be written without their knowing the amount of memory they will be allocated. The latter condition is most easily satisfied by the virtual memory technique described earlier, the former may also be satisfied as a result of the address translation mechanism used to implement virtual memory. Each task is constructed as though it owned the entire virtual address space, then the address is extended to include a task identifier for translation purposes.

It is obligatory in multi-user systems (where there are multiple tasks each belonging to a different user), and very desirable in single-user multi-tasking systems, to protect individual tasks from each other. If one task suffers an error which causes it to start writing randomly all over store, it will corrupt its own space but should be unable to corrupt memory being used by other tasks. The operating system must be able to access all task spaces to load data from disk when required.

Multi-user systems should make it impossible for a user to destroy another user's space even when the attempt is deliberate. This means, for instance, that a malicious user must not have access to the translation tables, otherwise he could alter them to switch another users space into his own and then corrupt it. Nor must he have access to another user's disk space, and so on.

These requirements lead to various privilege and protection mechanisms in systems, which usually revolve around a translation mechanism which keeps task spaces apart, and at least one privileged *supervisor* mode in the processor which alone is enabled to change translation tables and handle physical IO. Entry to supervisor mode forces transfer of control to a trusted area of code (which must also be protected from user modification), and this code can treat the source of the supervisor call with appropriate suspicion and check that the requested service is authorized.

Like virtual memory, support for multi-tasking is now mandatory on general-purpose computers above the PC level, and it is becoming important on PCs.

A RISC processor must therefore offer the necessary support if it is to compete in these market places. This results in additional complexity inside the processor.

1.3.4 Cache Memory

The factors which determine main memory access and cycle times are not the same as those which determine processor instruction rates, and a processor can easily become limited by the rate at which new instructions and data can be supplied by the memory. A cache memory is a small high speed memory which keeps copies of recently used memory items. Computer programs spend a lot of time in loops, and on the second and subsequent iterations of a loop the instructions will be found in the cache. This allows the processor to operate at cache speeds much of the time, rather than being held back to main memory speeds. The use of data items also frequently displays the property of locality (stacks being an obvious example), and data caches are often used.

The performance of a cache depends on its *hit ratio*, which is the proportion of all the cachable memory requests which turn out to be in the cache. This number depends on the size and organization of the cache, and also on the characteristics of the program which is being executed. For a detailed analysis of the dependence of cache performance on various design parameters see Smith (1982). Here we will describe just the main organizational features.

Starting from an uncached CPU (figure 15), there are two principal ways of attaching a cache memory. The cache may connect either to the virtual address bus or to the physical address bus (figure 16). In the former case the translation look-aside buffer (TLB) may perform the virtual to physical address translation in parallel with the cache look-up, so that the physical address is available immediately when a cache miss arises. With this cache arrangement care is needed on process switches or whenever the translation tables are modified, since either of these actions may cause entries in the cache to become out of date. A physically addressed cache avoids this problem, but requires a translated address, which may slow the cache access.

In both cases care is needed when physical memory is changed by an agent other than the processor attached to the cache, for instance a direct memory access from an intelligent peripheral controller. The most complex situation to handle arises in systems with multiple processors; we shall return to this later.

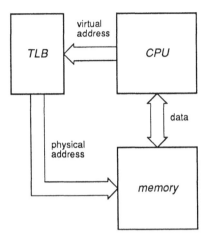

Figure 15: An uncached CPU

The cache must contain not only recently used data values, but also a *tag* which can be used to establish whether future references are to the same address. The tag may also contain information identifying the process in whose address space the entry resides (especially relevant to virtually addressed caches), and the privilege level that the processor was operating at when the data was fetched. It is not much use having protected supervisor code unless the cache can also prevent unauthorized access to it.

An important consideration when designing a cache is the algorithm which will be used to decide where to put new data in the cache. The simplest system is the *direct mapped* cache (figure 17), where the low address lines are used to determine a unique location. This scheme is very simple, and works well for large caches. The disadvantages are obvious; when a program loop, and the data it is working on, reside at addresses which have the same bottom bits, the program and data will contend for the same cache entry. A similar problem will arise when the program jumps back and forth between two sections which happen to share the same bottom few bits. These clashes become more problematical as the cache size is reduced.

An alternative organization is the *dual-set associative* cache (figure 18). Here the low address lines specify a pair of locations, and an entry may be placed in either. A single bit can record which of the two was last used, and the other is used this time. Alternatively, the selection may be random. Now a

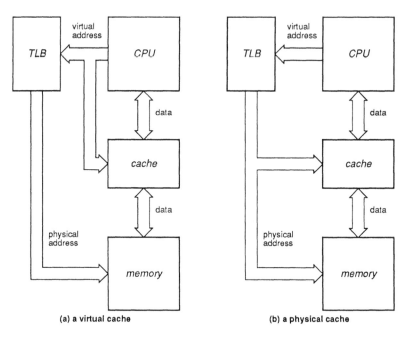

Figure 16: Two ways of addressing a cache memory

program must switch repeatedly between three uses of the same low order bits before contention arises. This is less likely than the two required to cause contention in the direct mapped cache.

The associativity may be increased to four-way, and beyond, but note that the algorithm for determining the least recently used (LRU) entry becomes much more complex.

The limiting case is the *fully associative* cache (figure 19). No low order bits are decoded now, and a data value may reside in any of the entries. Every tag must have its own comparator, so the tag is stored in content addressable memory (CAM) where a new address is compared simultaneously with the addresses of all the entries. If a match is found the corresponding data is sent to the CPU. The LRU algorithm to determine which entry to use next is complex to implement, and simpler alternatives are often used. Cycling round the entries in sequence is one possibility, or using them pseudo-randomly is

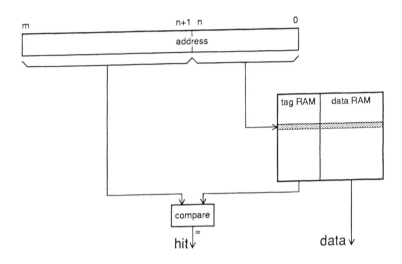

Figure 17: Direct mapped cache organization

another. The pathological cases where contention arises for a small subset of the entries are much less common than with the direct mapped cache, though, in the case of the pseudo-random replacement algorithm, they can be remarkably obscure when they do arise.

If a 32-bit CPU with 32-bit addressing has a cache with an address tag for every word, the tag store will be as large as the data store. Since it is the data which is really wanted, and the tag is overhead on the data, this seems less than optimal. There is no reason, however, why the data item stored with a tag should be restricted to a word. Programs usually use more than one word from a particular locality, and it usually makes little difference to the hit rate if four words are kept with every tag (and the hit ratio may be significantly improved if all four words are fetched in response to the first miss in a block). The tag overhead can thereby be reduced to one quarter of the data store, which seems much more reasonable. Large caches may have even larger lines associated with each tag. Figure 20 shows a fully associative cache with a quad-word line size. The conversion of the other cache architectures to quad-word lines is similar.

There are still more options to determine a complete cache design, for instance whether one cache should contain both instructions and data, or

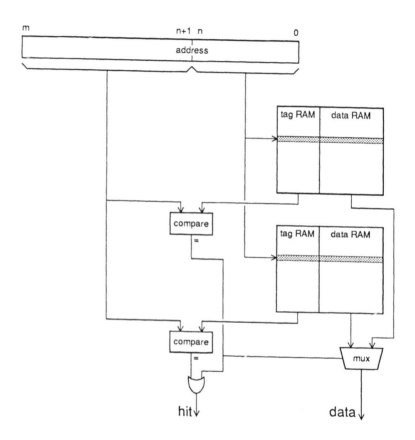

Figure 18: Dual-set associative cache organization

should a separate cache be used for each. A mixed cache is more flexible, but has less bandwidth than two separate caches.

The strategy to be used with data written by the CPU is important. It may always be sent back to main memory (a *write-through* strategy), or it may be written only into the cache (a *write-back* strategy). The write-through approach keeps main memory up to date, which means that any entry may be overwritten with new data without any problem. When an entry is to be overwritten in a write-back cache, the location to be overwritten must first be checked to see if it contains written data, in which case that data must be

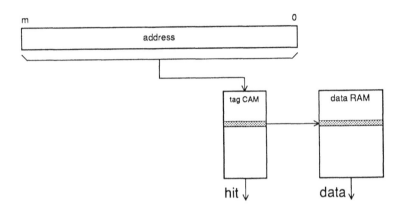

Figure 19: Fully associative cache organization

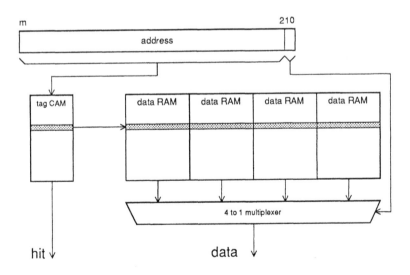

Figure 20: Fully associative cache with quad-word line

written back to main memory before the location may be re-used.

Whatever the write strategy, a *write buffer* may be used to allow the processor to continue without having to wait for a full main memory access. The data is written to the buffer, which subsequently passes it on to main memory, but in the meantime the processor continues with the following operations. The buffer could handle just one write operation at a time, in which case the processor will stall if a second write occurs before the first has completed, or it could contain a queue of write operations which will not cause a stall unless the queue becomes full.

Caches are extensively used in high performance computers, and are now increasingly being used in 32-bit microprocessor systems, including those based on RISC architectures. RISC processors tend to require very high instruction bandwidths, so cache design is important for good performance.

Cache Consistency

Caches on multiprocessors cause problems because the various processors can end up with inconsistent views of the contents of main memory. One processor can update a copy of a memory location which it is holding in its cache, and a second processor may then attempt to access the corresponding main memory location. A mechanism is required to ensure that the correct (updated) value is passed to the second processor.

Several mechanisms have been proposed to ensure consistent cache operation. For example:

- *Write-through with invalidation.* All writes are passed to main memory, and all caches watch the bus and invalidate entries with the same physical address. This solution causes a large amount of bus bandwidth to be used to support the writes to memory.

- *Software control.* Software is responsible for identifying data items which are shared, and placing them in areas of memory which are either uncached or use write-through with invalidation. This still uses considerable bandwidth, since the bus must be used for all stores which are potentially shared, whether currently shared or not.

- *Berkeley ownership protocol* (Katz et al, 1985). If a cache needs to update a block, it must claim ownership of it (unless it already owns it). It does this by broadcasting its intention to all the other caches, thereby causing them to invalidate their copies. Once ownership has been obtained, the cache may update its copy without any reference to the bus, eventually copying the entry back to main memory when the block is replaced.

Whilst a cache retains ownership, it must intercept any reads for the data from the other caches, and supply the data itself if it has been modified. It must therefore *snoop* on the bus, which means it must watch all bus transactions and compare them with its contents. Snooping requires an access to the cache tag store for each memory transaction, which might interfere with a processor access, so a second tag store may be used.

Other protocols have been suggested, based on the cache knowing when an entry is shared (i.e. another cache currently holds a copy of the item). Write-through is used for items marked as shared, and copy-back for unshared items. This is still an area for research, with new proposals arriving regularly and no concensus on the best solution.

The Bell Labs C Machine Stack Cache

A novel cache architecture was proposed by Ditzel and McLellan (1982). Here there are no explicit registers in the CPU; instead the top of the current stack is held in a cache which is multiported like a register bank. Operands near the top of the stack may be accessed as quickly and with as few instruction bits as on-chip registers.

The cache is implemented as a circular buffer of registers, and its size may be implementation dependent in a way which is transparent to the programmer. The cache is controlled by explicit instructions at procedure entry and return, and gives each procedure a new workspace area without unnecessary saving and restoring of registers. Entries are copied out into main memory only when the circular buffer overflows. (See the description of the AT&T CRISP microprocessor in chapter 3 for more details.)

The operand usage characteristics of high-level language compilers are well matched to the behaviour of this cache architecture, and it has found favour with some RISC designers. It is similar in principle to the register windows scheme used on the Berkeley RISC designs (see chapter 2).

1.3.5 Hardware Accelerators

The addition of special-purpose function units to accelerate particular classes of operation is now quite common. They are usually built to handle particular data types which the CPU datapath manipulates relatively inefficiently, such as floating-point operands, and vectors and arrays of integer or floating-point numbers. Central processors spend most of their time handling addresses, which are large integers, and it does not make sense to compromise the handling of these types in order to improve the handling of floating-point numbers. It is far better to have a separate piece of hardware which is purpose-built to handle floating-point, with the main processor performing the address calculations to find the values in memory. Since some users are not too sensitive to floating-point performance, it may make sense to offer the floating-point unit as an upgrade for those who need it. The floating-point operations may be performed slowly by a subroutine library, or by microcode on the integer datapath when the hardware is not attached.

The VAX-11/780 floating-point accelerator is an example of a hardware accelerator. It may be thought of as an additional function unit on the main datapath; it uses operands from the integer registers, and returns results to them. This approach is fine for the VAX, but VLSI processors do not allow external access to their datapaths. A different model is required.

Accelerators for VLSI CPUs are usually called *coprocessors*, and are typically VLSI devices themselves. There are several ways used to interface them to the main CPU:

- *Bus watching.* The coprocessor sits on the main memory bus and watches the instruction stream flowing into the main CPU. When a coprocessor instruction is fetched, it is executed by the coprocessor rather than the CPU. If the instruction involves transferring data to or from memory, the CPU may generate the address (using its integer registers) and the coprocessor controls the data bus. The coprocessor will have its own internal register set which must be saved on context switches, and transfers to CPU registers will be explicit.

  This system becomes more complex when the CPU has an on-chip instruction cache. The instruction stream is not visible on the main memory bus, and must be exposed to the coprocessor via a dedicated bus (which must carry the high bandwidth instruction stream).

- *Rebroadcasting.* Here the CPU has sole responsibility for recognizing coprocessor instructions, and when it finds one it retransmits it (possibly with some translation) along with a control signal which tells the coprocessor to execute it. The rebroadcast route may be down the system bus, or along a special-purpose bus. Operands may be sent along with the instruction, and the result returned to a CPU register, or the coprocessor may have separate registers.

Here a cache presents no special problems; only genuine coprocessor instructions need be retransmitted. The performance is likely to be lower than the first scheme because of the need for the CPU to identify relevant instructions before giving them to the coprocessor, whereas in the first scheme the CPU and coprocessor decode the instruction at the same time.

- *Memory mapping.* The coprocessor may be treated as a memory-mapped peripheral, with commands and data being fed to it by standard CPU instructions. This is the simplest and lowest performance solution.

RISC CPU designers cannot afford to ignore the importance of good floating-point performance unless they are happy to shut their design out of large market segments, and high performance coprocessor interfaces are quite complex. Good floating-point functionality is increasingly being included as part of the main CPU, often as a separate on-chip function unit. Otherwise, it is necessary to allow for an external accelerator.

1.3.6 Parallel Processing

A simple von Neumann machine fetches an instruction, decodes it, fetches operands from memory, operates on them, then writes back the result. It then fetches the next instruction. The clock which determines the rate of execution may be made faster until some component of the system stops working. The component which fails at the lowest clock rate determines the maximum possible performance of the machine, and is the system bottleneck. The obvious way to speed up the system is to overlap successive instructions to ensure that the limiting component is always busy, then to introduce some degree of parallelism into the CPU by duplicating the limiting function or functions. If the machine is still too slow, add another CPU!

Early RISC designs achieved their performance by exploiting the first of these approaches to the full, through the use of a technique called *pipelining*. Later designs have included parallel function units to a limited degree, and many are configured to enable efficient multiprocessing. We shall look at each of these approaches in turn. Bear in mind that they are not mutually exclusive; as was indicated in the description above, all may be brought to bear at once to maximize system performance.

Pipelining

A simple form of parallelism is achieved by overlapping successive instructions. A typical instruction can be split into several phases, for instance:

(1)   Instruction fetch

(2)   Instruction decode

(3)   Operand fetch

(4)   ALU operation

(5)   Write result

Since each phase uses a different section of the processor, it is possible to overlap instructions one phase apart, so that although the instruction takes five phases to complete, the processor completes one instruction every phase (see figure 21).

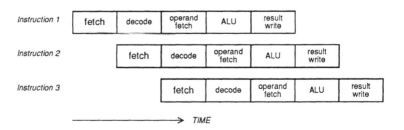

Figure 21: Instruction pipelining

Note that problems can arise, for instance if one of the operands for instruction 2 is the result of instruction 1. Instruction 2 will get the wrong value from the register, which won't be updated to the correct value until the next phase. Various solutions to this *data race hazard* are possible:

•   It could be left to the compiler not to produce code which displays this problem, by careful reordering of instructions and using no-ops where necessary (*software pipeline management*).

- A hardware interlock could detect the problem and delay instruction 2 until the result of instruction 1 has been written (*pipeline stalling*, see figure 22).

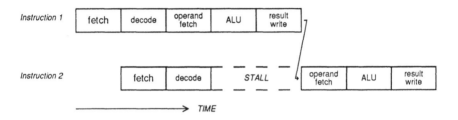

Figure 22: Pipeline stall solution to data race hazards.

- A hardware interlock could detect the problem and forward the result of instruction 1 directly to the operand register for instruction 2, without delaying it (*register forwarding*, see figure 23).

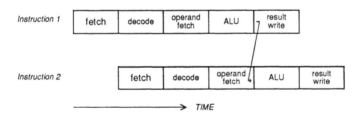

Figure 23: Register forwarding solution to race hazards.

The most serious problem for the pipeline arises when there is a branch instruction. It appears at first sight that all the instruction fetches following the branch were from the wrong address. They should be discarded and the pipeline refilled from the branch target address. A more efficient alternative is to execute these instructions anyway, and rely on the compiler to position the branch so that it happens at the correct time. Note that now the instructions do not execute in the obvious sequence.

The *delayed branch* solution to the pipeline break problem has been used for some time in microcode, and is now used by most RISC processors (which usually are deeply pipelined). It is often possible for the compiler to find one instruction to move after the branch, and not uncommon for two to be possible when the architecture allows.

Parallel Function Units

When more performance is required from a processor and the limiting component is the ALU, one option is to build multiple parallel arithmetic units and to use these to operate on several instructions concurrently. The Control Data 6600 (Thornton, 1964) had a register set connected to ten functional units (a branch calculator, two incrementers, a boolean logic unit, a shifter, two multipliers, one divider, an integer adder and a floating-point adder). By despatching instructions until no function unit is free, or the same register is to be used for the result of two different instructions, substantial concurrency can be achieved.

Extensive interlocks are required to account for cases where the result of one operation is an operand of a second operation, but none the less the speed up is significant compared with dispatching one operation at a time. The interlocks are implemented in a generalized queue and reservation scheme called the *scoreboard*, where a record is maintained of the usage state of each register, functional unit, and interconnecting bus. A new instruction causes a new record to be added into the scoreboard, which delays execution if necessary, but does not delay the issue of subsequent instructions.

RISC architecture is rapidly approaching the point where the ALU is the system bottleneck. Nearly all instructions take only one cycle, and VLSI allows the inclusion of caches large and fast enough to keep the CPU supplied with work. Parallel function units represent one of the few ways of improving uniprocessor performance from this position, other than just waiting for better semiconductor processing technology. Some RISC microprocessors already incorporate multiple function units and register scoreboards.

Multiprocessors

Once everything possible has been done to maximize the performance of a single processor, the only course left open to higher performance is to use more than one CPU. Indeed, the optimal strategy may be to develop uniprocessors only to the point of best price/performance, then use multiple CPUs for additional processing power, rather than aiming for peak uniprocessor performance. The difficulty is that a multiple CPU system is no longer a von Neumann computer, and completely new operating systems and languages are required to make efficient use of the parallel capabilities of the machine for general-purpose applications. The general industry trend is therefore towards maximizing uniprocessor performance first, since uniprocessor software issues are well understood, and building multiprocessors second.

When multiprocessor software systems are better established, the emphasis may shift. It will be possible to build several powerful CPUs on one chip. Finding the optimum trade-off point, of the power of a single CPU against the number that will fit on one chip, will be an additional challenge for the processor architect.

As general-purpose RISC multiprocessors are still in the research phase, we will leave this interesting subject for now. It will recur prominently in the final chapter when we speculate on the future of RISC, and also in chapter 3 in the section on the Inmos transputer, which is the only commercial VLSI CPU which has been designed specifically to support multiple CPU, fine-grain, parallel processing.

1.4 SOFTWARE CONSIDERATIONS

Sometimes new software technology drives CPU design, and sometimes the reverse is true. RISC processors appear to have forced the development of efficient register allocation algorithms, whereas the importance of procedures to most high-level language compilers has caused CPU designers to introduce special instructions and register organizations to support procedure calls and returns. The requirement for dynamic type checking in some AI languages has caused CPUs with hardware support for this to be developed (Pendleton et al, 1986).

Operating systems have also developed interactively with CPU hardware. Multi-tasking and virtual memory have already been mentioned; these both require extensive operating system software support in addition to the hardware support described earlier.

It is certainly a mistake to design a CPU without the close participation of operating system and compiler experts.

1.4.1 Operating Systems

A RISC CPU must support an operating system efficiently, and since building an operating system from scratch is a very expensive activity, it is usually desirable to support an existing one. The standard operating system in the market segment in which most RISC CPUs are used is UNIX, often with a window system and possibly a distributed filing system. UNIX at present supports multi-tasking with heavyweight process switches; process switches are relatively infrequent and include a complete change of processor and address translation system context. There is a trend in operating system design towards lightweight processes, with multiple threads of control in each process, where a process switch involves changing only the processor context (with much less frequent changes of the address translation context). This approach is being applied to new versions of UNIX (see, for instance, Rashid et al, 1987) as well as to completely new operating systems.

A consequence of switching to an operating system with support for lightweight processes is that the speed of processor context switches becomes more important, since these will happen more often. This might be an argument against large register files, since although these support procedure calls well, they lead to relatively slow processor context switching.

The operating system requirements need very close examination in the case of multiprocessors, to check what level of cache and translation buffer coherency is required, and what sort of interlocks are needed to control shared resources. Multiprocessor operating system design is still in its infancy, but developing rapidly. New ideas arise often, and these frequently place new requirements on the design of the CPU hardware.

1.4.2 High-Level Languages

The RISC movement has from the outset depended upon the compiler writer to bridge the increased semantic gap between the instruction set and high-level languages. Some RISC instruction sets impose significant extra loads on the code generator, by requiring it to manage non-interlocked pipelines. Most of them require good register allocation if performance is not to be thrown away, and so on.

High-level languages are more than abstract macro-assemblers; they allow the programmer to use mental pictures of what the program is doing which are totally unrelated to the underlying hardware. The languages assume the support of all manner of data structures, procedure calling mechanisms, multi-way branches, and so on, and the compiler must cause these to happen as expected. If the instruction set does not support a particular model efficiently, it will have to be done inefficiently.

Most compilers assume the existence of some sort of stack for the support of recursive algorithms, and for some it matters whether the stack grows up or down memory. If the language wants the stack to grow up memory but the CPU stack instructions all assume it grows down memory, then the compiler cannot use those instructions. It will have to synthesize the stack out of other instructions as best it can. The unusable stack instructions are a waste of functionality (for this language).

One area where RISC CPUs have forced compiler technology to improve is in register allocation. CISC processors support memory to memory operations, so registers tend to be used to hold pointers to data structures, and the data values themselves reside in memory. RISC CPUs typically have load/store architectures, so data values must be moved into a register before they can be used. This causes contention for register usage, and a high cost in throughput if incorrectly handled. Algorithms have been developed (Chaitin, 1982) which enable the compiler to make good decisions about which values should be held in registers, and which registers may be re-used. These algorithms postpone allocation until late in the compilation process, when a lot of information about usage has been gathered.

RISC CPUs also introduced the idea of register windows to reduce data traffic on procedure call and return, and these require new management techniques (Tamir, 1983). The software management of non-interlocked pipelines is another area where RISC CPUs have forced compiler development (Gross, 1983).

Interestingly, though these compiler developments were forced by the new architectural features of RISC CPUs, some (particularly those relating to efficient register usage) have also been found to improve the performance of compilers for CISC processors. Earlier work had already established that restricting compilers to carefully chosen subsets of complex instruction sets had little impact on performance (eg Lunde, 1977). This supports the view that a large complex instruction set is not the best interface between the hardware and the compiler.

1.5 THE DESIGN PROCESS

The sequence of activities which transform an architectural idea into a product is complex and beyond the scope of this book, but some observations may be made which will assist in putting the RISC movement into context.

1.5.1 Levels of Knowledge

The computer industry relies upon a broad structure of knowledge and expertise, and this structure can be split into many interdependent levels of knowledge. Figure 24 illustrates this view. To work in any level you must have a degree of understanding of neighbouring levels, but you may get by with no knowledge of remote levels. Technical people tend to specialize in one or two layers of this model, knowing only as much about neighbouring levels as is necessary to work at their specialist level.

The independence of the layers is enhanced by well defined interfaces. A language definition should free an operating system programmer from any requirement to understand the workings of a compiler. Likewise, a logic designer should be able to design in gates without reference to how the gates are constructed or what the underlying semiconductor process is.

Entire companies operate successfully without covering the complete range of activities. For instance semiconductor manufacturers need only think about instruction sets (and then only if they wish to build microprocessors) and below, and computer systems houses may take standard chips and not worry about how they are made (though this is much less often the case than it was a few years ago).

The significance of this view of the structure of computer related knowledge is that while most practitioners work within their level, major advances often come from moving the interfaces between these levels. The RISC philosophy is an example of this, where functionality is moved out of the hardware instruction set up into the software code generator. This can only be achieved by someone who understands both sides of the interface thoroughly.

The basic aim of the CISC designer is to try to get as close as possible to the ultimate CISC goal where every high-level language statement is mapped into a single CPU instruction; CISC designers take this philosophy as given, and apply their efforts accordingly. The real breakthrough of the RISC movement is to dare to question this assumption.

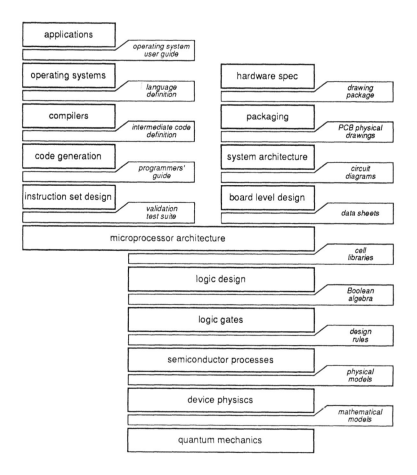

Figure 24: Levels of knowledge in computer science

1.5.2 Balanced Systems

An installed computer system represents an investment in a number of
different functions which are expected to work together to deliver a desired
result. In any given application, one part of the system will be the factor
which limits the performance of the entire system. All other system functions
will be working below their potential, so they could have been designed to a
lower specification (at lower cost) without affecting the performance of the

system in this application. Money has been wasted.

If the bottleneck is different for different applications or users, then the design compromise may be correct overall, and economies of scale dictate that it is cheaper to have one design cover a range of uses than for every installation to be uniquely matched to its load. If, however, the same factor is always the limiting one, then the system is unbalanced.

So how is a system designed to be balanced? Amdahl suggested two rules of thumb whilst designing the IBM System/360 (Siewiorek, Bell and Newell, 1982, p. 46). The first relates the physical memory size to the processor speed; there should be one byte of main memory for every instruction per second. The second relates I/O bandwidth to processor speed; there should be one bit per second of I/O for every instruction per second. A 1 MIPS (Million Instruction Per Second) processor should have 1 Mbyte of memory and 1 Mbit/s of I/O.

These rules should probably be applied to orders of magnitude rather than taken as exact. They were proposed at a time when computers were departmental mainframes, and the aim was to keep all parts of such a machine fully loaded. In such a machine there is a queue of programs awaiting execution, each of which will be loaded into memory (using I/O bandwidth), occupy some amount of memory, consume some number of CPU cycles and some I/O for data files, and produce results which must be stored somewhere for future reference (more I/O).

It is not clear that the same rules apply to personal computers and workstations. PCs lie idle much of the time, but when activated are required to produce a result as soon as possible. Their load is very unbalanced, which makes it much less easy to define what is meant by balanced system hardware. The trend in personal computing is, however, towards more sophisticated operating systems, with multi-tasking to support multiple foreground tasks and background activities. Such an environment offers a more balanced load, and Amdahl's rules are likely to become increasingly relevant as this trend accelerates.

It should be noted here that the CPU and memory technologies are roughly keeping pace with each other in terms of balance, so VLSI CPU speeds are increasing in line with VLSI memory sizes. It would not seem, however, that peripheral technologies (especially disks) are increasing in bandwidth at a sufficient rate to maintain the balance of low cost personal systems, so some architectural adjustments seem likely in the near future to compensate. The extensive use of additional main memory for keeping a cache of disk files would reduce the I/O requirement in a stand alone system, or several machines could jointly bear the cost of an expensive high-performance disk if a suitable high speed LAN were economically available. High capacity battery backed-up VLSI memory systems are also becoming economic, and have much higher bandwidth than disks.

The relevance of system balance to the RISC movement is that it points to the necessity of considering the performance of the system as a whole as opposed to putting effort solely into raising processor performance. A very fast processor will simply unbalance the system with little benefit to the user unless the memory size and I/O performance increase proportionately. A semiconductor manufacturer can leave these as problems for the user of its chips, but a computer system manufacturer must solve all the problems at once.

1.5.3 Market Success

The computer market is technology driven. Users expect better price/performance year by year, and are attracted by new systems which claim greater performance than last year's models. However they are not seduced by claims alone, and it will rapidly become known if a new machine does not deliver because of some bottleneck in the system.

The rate of progress is such that a new machine can only be expected to maintain a technical lead for a short time. If during that time it is well marketed, a user base can be established which will generate the applications for the second phase of product life. During this second (typically much longer) phase the product will not have a perceived technical lead. It will depend for sales on its application base and good support and marketing.

If a new product is compatible in some way with an existing machine which has already established an application base, the new product has a very substantial head start over competing products which can't exploit existing software. A RISC system is unlikely to be object code compatible with an existing system, but other forms of compatibility are useful and possible. The most usual is the high-level language compatibility which results from implementing industry standard compilers and operating systems such as UNIX. Other approaches to compatibility include software emulation of an older processor, or even the inclusion of the older processor as an optional hardware extension. The former is effective when the older processor is an order of magnitude less powerful than the RISC, in which case the emulation may be as fast as the original machine. The latter is most often used as a temporary bridge while software is moved to the new architecture.

If not one but a range of machines can be developed, each balanced at a different level of performance, and offering good compatibility between models, users will feel more confident that they will be able to migrate up the range as their computing needs increase. This will improve the chances of market success, which must be the major goal of any design process. Any range of machines which covers a significant spread in performance must embody different CPU technologies to match each performance point economically, so the commercial RISC (or CISC) designer should aim for an architecture which allows efficient implementation at several performance points.

1.6 THE VAX-11/780

The VAX-11/780 architecture was introduced in 1978, and represents the pinnacle of the pre-RISC evolutionary development. It embodies all the architectural features which are required to support high performance multi-user operating systems, and incorporates a very rich and powerful instruction set. Partly because of its well balanced power, but equally because of its commercial success, the VAX-11/780 is the standard against which all new systems are compared, even a decade after its introduction.

We will look in some detail into the architecture of the VAX-11/780, as it symbolizes the base from which the RISC movement was launched. (See Strecker, 1978, and Digital Equipment Corporation, 1980 for more detail.)

1.6.1 Architecture

The VAX application programmer sees a virtual address area of 4 Gbytes (figure 25), of which he can use 1 Gbyte. 1 Gbyte is used by the system to store information particular to the application, and a third Gbyte contains general system information. The fourth Gbyte is reserved for future use.

The processor contains 16 32-bit registers which are visible to the programmer, four of which have special functions (figure 26). The remainder may be used to hold address or data values. The program status longword (figure 27) contains the remaining user accessible state, including access mode bits which indicate the privilege level of the current activity. There are four copies of the stack pointer (register 14) to support independent stacks for each of the privilege levels.

Data Types

The instruction set supports a wide range of data types explicitly. All may be stored on any byte boundary; the bit field type may be stored at any bit position in memory.

- 8-bit (B), 16-bit (W), 32-bit (L) and 64-bit (Q) integers, which may be considered to be unsigned or 2's complement signed numbers.

- 32-bit (F) and 64-bit (D) floating-point numbers.

- 0 to 65535 byte character strings.

- 0 to 31 digit packed decimal or ASCII decimal strings.

- 0- to 32-bit bit fields, which may be unsigned or 2's complement signed.

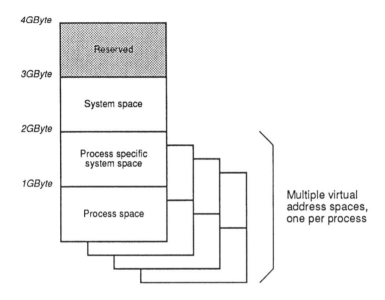

Figure 25: The VAX-11/780 virtual address space
(Adapted from Digital Equipment Corporation, 1980)

Addressing Modes

An instruction may specify the address of an operand in the following ways:

• The operand may be held in a register.	Rn
• Its address may be in a register.	(Rn)
• The register may be auto pre-decremented	(Rn-)
• or auto post-incremented	(Rn)+
• and possibly indirected.	((Rn)+)
• The register may be added to a displacement	(Rn+D)
• and possibly indirected	((Rn+D))
• or an immediate value may be used.	I

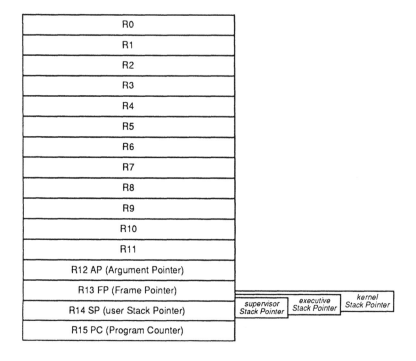

Figure 26: The VAX-11/780 register organization

All of the addressing modes except Rn and I may be modified by adding a scaled index register Rx to the operand address. Note that Rn may be the PC, so program relative addressing modes are well supported.

Instruct⁻ Set

The VA 1' 780 supports all the usual logical and arithmetic operations on relevant ¸a'¸ types, instructions to convert between the types, and branching instructions. In addition there are some very complex instructions, such as procedure call and return, case, save and load process context, add and delete item from a linked list. The full instruction set is listed in table 2.

It is clear that the VAX-11/780 has a very extensive instruction set. It comes very close to the abstract machine expected by the high-level language

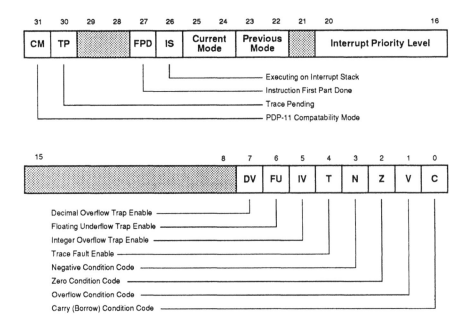

Figure 27: The VAX-11/780 program status longword
(Adapted from Digital Equipment Corporation, 1980)

compiler writer, especially for languages like C, Pascal and Modula 2. It provides all the necessary facilities for writing protected operating systems, and has allowed many different implementations of both faster machines and cheaper machines, including VLSI implementations of the CPU. It also has dense code.

It should be noted that the procedure call instruction, in addition to its obvious role in saving code space, enforces a standard procedure call mechanism. This has many benefits, which include the simplification of inter-language calls. Systems without such an instruction allow each compiler writer to adopt a procedure call which suits only his language, making inter-language calls very difficult. Clearly the compiler writers could get together to agree a call standard to overcome this problem, but embedding the standard in the hardware instruction set is a good way of forcing the agreement!

Table 2. The VAX-11/780 instruction set

Integer and Floating-Point Logical Instructions

MOV/MNEG	*Move/Move Negated (B, W, L, F, D, Q)*
MCOM	*Move Complemented (B, W, L)*
MOVZ	*Move Zero-Extended (B→W, B→L, W→L)*
CLR	*Clear (B, W, L, F, D, Q)*
CVT	*Convert (B, W, L, F, D)→(B, W, L, F, D)*
CVTR-L	*Convert Rounded (F, D) to Longword*
CMP	*Compare (B, W, L, F, D)*
TST	*Test (B, W, L, F, D)*
BIS/C	*Bit Set/Clear (B, W, L) 2- and 3-Operand*
BIT	*Bit Test (B, W, L)*
XOR	*Exclusive OR (B, W, L) 2- and 3-Operand*
ROTL	*Rotate Longword*
PUSHL	*Push Longword*

Integer and Floating-Point Arithmetic Instructions

INC,DEC	*Increment/Decrement (B, W, L)*
ASH	*Arithmetic Shift (L, Q)*
ADD	*Add (B, W, L, F, D) 2- and 3- Operand*
ADWC	*Add with carry*
ADAWI	*Add Aligned Word Interlocked*
SUB	*Subtract (B, W, L, F, D) 2- and 3-Operand*
SBWC	*Subtract with Carry*
MUL/DIV	*Multiply/Divide (B, W, L, F, D) 2- and 3-Operand*
EMUL/EDIV	*Extended Multiply/Divide*
EMOD	*Extended Modulus (F, D)*
POLY	*Polynomial Evaluation (F, D)*

(continued)

Table 2. (continued)

Index Instruction

INDEX   *Compute Index*

Packed Decimal Instructions

MOVP	*Move Packed*
CMPP	*Compare Packed 3- and 4-Operand*
ASHP	*Arithmetic Shift Round and Packed*
ADDP/SUBP	*Add/Subtract Packed 4- and 6-Operand*
MULP/DIVP	*Multiply/Divide Packed*
CVT	*Convert Long/Trailing/Separate to/from Packed*
EDITPC	*Edit Packed to Character String*

Character String Instructions

MOVC	*Move Character 3- and 5-Operand*
MOVT(U)C	*Move Translated (Unit) Characters*
CMPC	*Compare Characters 3- and 5-Operand*
LOCC/SKPC	*Locate/Skip Character*
SCANC/MATCHC	*Scan/Match Characters*

Processor State Instructions

PUSHR/POPR	*Push/Pop Registers to/from Stack*
MOVPSL	*Move from Processor Status Longword*
BISPSW/BICPSW	*Bit Set/Clear Processor Status Word*

(continued)

Table 2. (continued)

___

Variable-Length Bit Field Instructions

EXTV	*Extract Field*
EXTZV	*Extract Zero-Extended Field*
INSV	*Insert Field*
CMPV	*Compare Field*
CMPZV	*Compare Zero-Extended Field*
FFS	*Find First Set*
FFC	*Find First Clear*

Branch on Bit Instructions

BLB	*Branch on Low B (S, Cl)*
BB	*Branch on Bit (S, Cl)*
BBS	*Branch on Bit Set and (S, Cl) Bit*
BBC	*Branch on Bit Clear and (Set, Clear) Bit*
BBSSI	*Branch on Bit Set and Set Bit Interlocked*
BBCCI	*Branch on Bit Clear and Clear Bit Interlocked*

Queue Instructions

INSQUE	*Insert Entry in Queue*
REMQUE	*Remove Entry from Queue*

Address Manipulation Instructions

MOVA	*Move Address (B, W, L, F, D, Q)*
PUSHA	*Push Address (B, W, L, F, Q, D) on Stack*

___

(continued)

Table 2. (continued)

Unconditional Branch and Jump Instructions

BR	*Branch with (B, W) Displacement*
JMP	*Jump*

Branch on Condition Code

BLSS(U)	*Less Than (Unsigned)*
BLEQ(U)	*Less Than or Equal (Unsigned)*
BEQL/BNEQ	*Equal/Not Equal*
BGTR(U)	*Greater Than (Unsigned)*
BGEQ(U)	*Greater Than or Equal (Unsigned)*
BVS/BVC	*Overflow Set/Clear*

Loop and Case Branch

ACB	*Add, Compare and Branch (B, W, L, F, D)*
AOBL	*Add One and Branch Less Than (or Equal)*
SOBG	*Subtract One and Branch Greater Than (or Equal)*
CASE	*Case on (B, W, L)*

Subroutine Call and Return Instructions

BSB	*Branch to Subroutine with (B, W) Displacement*
JSB	*Jump to Subroutine*
RSB	*Return from Subroutine*

(continued)

Table 2. (continued)

Procedure Call and Return Instructions

CALLG	*Call Procedure with General Argument List*
CALLS	*Call Procedure with Stack Argument List*
RET	*Return from Procedure*

Access Mode Instructions

CHM	*Change Mode to (Kernel, Executive, Supervisor, User)*
REI	*Return from Exception or Interrupt*
PROBER	*Probe Read*
PROBEW	*Probe Write*

Privileged Processor Register Control Instructions

SVPCTX	*Save Process Context*
LDPCTX	*Load Process Context*
MTPR	*Move to Process Register*
MFPR	*Move from Process Register*

Special Function Instructions

CRC	*Cyclic Redundancy Check*
BPT	*Breakpoint Fault*
XFC	*Extended Function Call*
NOP	*No Operation*
HALT	*Halt*

(Source: adapted from Strecker, 1978.)

1.6.2 Organization and Implementation

The VAX-11/780 CPU is built of Schottky TTL components, and uses microcode. The microcode is partly fixed and partly writeable. The writeable section is used to implement some instructions and to patch others, and to enable instructions to be built for diagnostic purposes. The floating-point accelerator (FPA) is an optional CPU upgrade which increases the performance (but not the functionality) of the CPU when performing floating-point instructions. Note that the floating-point operands are held in the main CPU registers; the FPA does not contain user state, it merely enables the floating-point functions to be performed faster.

The CPU contains an 8 byte instruction prefetch buffer, and an 8 Kbyte cache memory for data. The cache is 2-way set associative with an 8 byte block size, and uses a write-through strategy with a write buffer. The replacement algorithm selects randomly between the two possible block locations. The cache watches the memory bus (the SBI) and takes copies of any I/O transfers into locations which are currently in the cache, thereby ensuring cache consistency.

The virtual to physical address translation is performed via page tables in virtual memory, and a 128-entry translation cache holds recently used page table entries. 64 of these are system entries and 64 are current process entries. The translation mechanism for system virtual addresses is shown in figure 28.

The translation mechanism for user process addresses is similar, except that the page table entry address which is obtained by adding the virtual page number to the process base register is now a system virtual address, and not a physical address. This system virtual address must be translated to locate the page table entry, so up to three memory references may be required to access a data item. Fortunately, the translation cache has a high hit rate, so that the address translation is usually directly available and only the memory reference to access the data item is performed.

The CPU communicates with memory and I/O subsystems via a bus called the Synchronous Backplane Interconnect (SBI), which transfers a 32-bit data item or a 30-bit address every 200ns. Two data items may be supplied for each address, giving a potential bandwidth of 13.3 Mbytes/sec. During a CPU read the bus is acquired by the CPU for sending the address and then released. The memory controllers will re-acquire the bus for transmission of the data when it is ready, and in the meantime the bus is free for other use. The memory controllers buffer up to four requests at a time. The memory

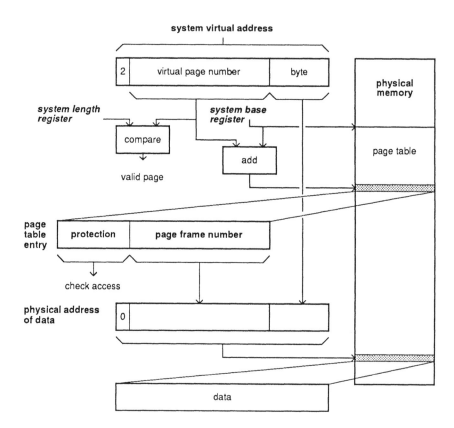

Figure 28: The VAX-11/780 translation mechanism for system addresses
(Adapted from Digital Equipment Corporation, 1980)

itself is modular, with different size configurations being available. The
memory system incorporates error checking and correction (ECC) logic. The
I/O is performed via buffered adaptors to the older Unibus and Massbus
interfaces, as used on the PDP/11.

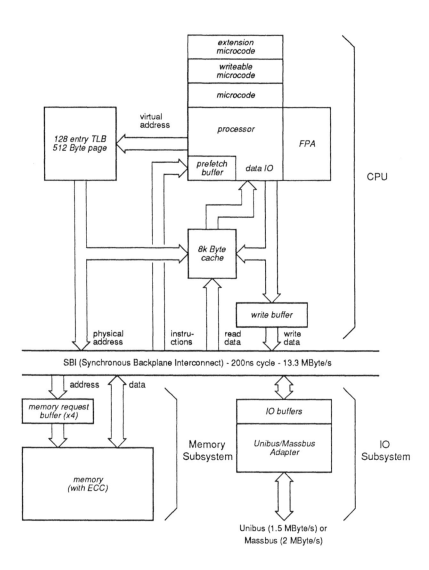

Figure 29: The VAX-11/780 system organization
(Adapted from Digital Equipment Corporation, 1980)

1.7 THE REDUCED INSTRUCTION SET COMPUTER

The evolution of computer architecture has been marked by the introduction of increasingly sophisticated solutions to the fundamental problem of maximizing performance whilst minimizing cost. A major cost in operating a computer system is that associated with generating programs, and computer hardware designers have addressed this issue by building increasingly complex instruction sets which move the hardware closer to the programmer's model of what he wants the machine to do. This approach was epitomized by the VAX-11/780 which was introduced in 1978, and included (for example) a single instruction procedure call, and single instructions for adding or deleting entries in doubly linked lists. The VAX instruction set is very powerful and this has brought a number of benefits; for instance, the procedure call instruction has had the very desirable effect of standardizing procedure entry mechanisms across different languages.

Against this background of rising complexity, the concept of the Reduced Instruction Set Computer (RISC) was introduced in 1980. The assumptions which led to complex instruction sets had been under question for some years, but it was the clear proclamation of the RISC approach by Patterson and Ditzel (1980) which turned the idea into a major trend.

Their argument in favour of RISCs related mainly to VLSI implementations of processors, and centred on three points:

- *Chip area.* A simple chip requires less logic area, so, as the number of transistors which can be put on a chip increases, it will be possible to get a complete RISC processor on a chip earlier than a CISC processor. This applied to silicon in 1980, and applies to gallium arsenide now (1988). Once transistor densities are such that both architectures are feasible, it is argued that a RISC will always leave more room for performance enhancing features such as caches.

- *Design Time.* CISC processors take much longer to design and debug than RISCs, so the development costs are higher, and it is much harder to keep pace with semiconductor process developments. A shorter development cycle should enable a RISC to be closer matched to current processes.

- *Performance.* It is a general property of digital systems that smaller logic structures go faster. Average gate fan-outs are reduced, and signal paths are shorter. A RISC processor should, all other things being equal, have a shorter cycle time because it is smaller and simpler. If adding an instruction slows the cycle without having a corresponding gain at a higher level, it should be left out. There is little evidence that most complex instructions can justify their inclusion against this criterion.

Following this declaration of the principle, both Berkeley and Stanford Universities built research prototypes of VLSI RISC processors. These demonstrated that the principle could be put into practice, and the delivered performance was good, though neither university had the benefit of the finely tuned manufacturing processes used by commercial microprocessor suppliers. Both these processors embodied the RISC philosophy, but they muddied the waters somewhat by including other new and interesting features which were peripheral to the RISC issue, but which have nevertheless become confused with it as a result. The Berkeley RISC (Katevenis, 1985) had a large windowed register file, and the Stanford MIPS (Hennessy et al, 1981) had an execution pipeline without hardware interlocks. Both features are worthy of research, but neither is central to the RISC approach. (It could be argued that the Berkeley register file illustrated a use of the silicon which is left free as a result of the RISC design, but it is not the only use to which that free area can be put.)

The academic lead was followed rapidly by commercial organizations keen to gain an edge from the improved performance offered by the RISC approach. The earliest examples were not VLSI implementations, but used standard logic parts. They served to demonstrate that the philosophy could be used to make complete systems.

1986, '87 and '88 have seen the introduction of several commercial 32 bit RISC microprocessors. These have included designs from companies new to the microprocessor business (eg Acorn, MIPS, Sun) and some from the established suppliers. The arrival of the RISC approach to computer design has caused large cracks to appear in the monopolies of the major CISC microprocessor manufacturers. This has coincided with a period of rapid technological advance which will further stress the market positions of those companies, creating an opportunity for new contenders to gain a hold.

Current forecasts suggest that the processing power of 32-bit microprocessors will double every year for the next few years from the current (1988) 10 to 15 MIPS level, and RISC architectures are likely to lead this advance because of their shorter design cycles.

1.7.1 Concluding Remarks

In this chapter we looked at basic computer architecture from a historical perspective, and introduced the significant architectural developments which have had lasting value in improving the performance and/or cost-effectiveness of general-purpose computers. We observed the rapid progress in the electronic technologies with which computers are constructed. All these factors contributed to the evolution of the Complex Instruction Set Computer, as epitomized by the VAX-11/780. These machines have very rich instruction sets, with instructions for high-level functions like procedure call and return, in an attempt to bring the instruction set as close as possible in concept to the high-level languages which would be used to program the machines. The VAX-11/780 was the first member of one of the most successful ranges of computers ever built, where the success may be measured either in commercial terms or in terms of user satisfaction. The architecture supports modern operating systems efficiently, it is well balanced, runs high-level languages well, and it is very flexible - a true general-purpose machine. It has also allowed other implementations of the same architecture which span a very wide range of price and performance points, both above and below the original 11/780.

Then, in 1980, it was suggested that the basic approach of the designs of machines like the VAX was wrong, or at least sub-optimal for VLSI implementations. Instead, the Reduced Instruction Set Computer was proposed, which discarded many years of instruction set evolution in favour of a much more basic set of instructions, optimized for efficiency of the hardware implementation. The proposal was supported by careful argument and, later, by prototype implementations of VLSI processors. In the next chapter we shall look at these pioneering RISC CPUs, and examine the RISC versus CISC arguments as presented by their designers.

Gradually, an increasing proportion of commercial computer manufacturers became sufficiently convinced by the evidence that they made significant investments in their own RISC CPU development programmes. Some of the resulting processors are close relatives of their research antecedents, others take only the bare RISC philosophy and apply it in different directions. Some mix a bit of the RISC approach with a bit of the CISC approach, and others have arrived near the RISC position completely independently by different routes. In chapter 3 we will look at the resulting diversity of commercially developed RISC and RISC-like VLSI CPUs.

References

Chaitin, G. J. (1982). Register Allocation and Spilling via Graph Coloring. Proceedings of SIGPLAN 82 Symposium on Compiler Construction, SIGPLAN Not. 17, no. 6, pp. 98-105.

Digital Equipment Corporation (1980). VAX Hardware Handbook.

Ditzel, D. R. and McLellan, H. R., (1982). "Register Allocation for Free: The C Machine Stack Cache," Proceedings of the Symposium on Architectural Support for Programming Languages and Operating Systems, pp. 48-54.

Gross, T. (1983). Code Optimization of Pipeline Constraints. PhD Dissertation, Stanford University. Stanford University CSL Technical Report 83-255.

Hennessy, J. L., Jouppi, N., Baskett, F., and Gill, J. (1981). "MIPS: A VLSI Processor Architecture," Proceedings of the CMU Conference on VLSI Systems and Computations. Computer Science Press, Rockville, MD, pp. 337-346.

Hitachi (1987). Hitachi Technology Transfer, Hitachi Electronic Components (UK) Limited, 21 Upton Road, Watford, Herts.

Katevenis, M. G. H. (1985). Reduced Instruction Set Computer Architectures for VLSI. MIT Press, Cambridge, MA.

Katz, R. H., Eggers, S. J., Wood, D. A., Perkins, C. L. and Sheldon, R. G., (1985). "Implementing a Cache Consistency Protocol," Proceedings of the 12th International Symposium on Computer Architecture, ACM SIGARCH, Boston MA.

Kilburn, T., Edwards, D. B. G., Lanigan, M. J. and Sumner, F. H. (1962). One-Level Storage System, IRE Transactions, EC-11, 2, pp. 223-235. (Reproduced in Siewiorek, Bell and Newell, 1982, pp. 135-148.)

Lunde, A. (1977). Empirical Evaluation of Some Features of Instruction Set Processor Architecture, Communications of the ACM, 20, no. 3, pp. 143-153.

Patterson, D. A. and Ditzel, D. R. (1980). The Case for the Reduced Instruction Set Computer, Computer Architecture News, 8, no. 6, pp. 25-33.

Patterson, D. A. and Sequin, C. H. (1981). "RISC 1: A Reduced Instruction Set VLSI Computer," Proceedings of the 8th Annual Symposium on Computer Architecture, pp. 443-457.

Pendleton, J. M., Kong, S. I., Brown, E. W., Dunlap, F., Marino, C., Ungar, D. M., Patterson, D. A. and Hodges, D. A. (1986). A 32-bit Microprocessor for Smalltalk, IEEE Journal of Solid-State Circuits, SC-21, no. 5, pp. 741-749.

Radin, G. (1983). The 801 Minicomputer, IBM Journal of Research and Development, 27, no. 3, pp. 237-246.

Rashid, R., Tevanian, A., Young, M., Golub, D., Baron, R., Black, D., Bolosky, W. and Chew, J. (1987). "Machine-Independent Virtual Memory Management for Paged Uniprocessor and Multiprocessor Architectures," Proceedings of the 2nd International Conference on Architectural Support for Programming Languages and Operating Systems, Palo Alto, pp. 31-39.

Siewiorek, D. P., Bell C. G. and Newell, A. (1982). Computer Structures: Principles and Examples, McGraw Hill.

Smith, A. J. (1982). Cache Memories, ACM Computing Surveys, 14, no. 3, pp. 473-530.

Stevens, W. Y. (1964). The Structure of System/360, IBM Systems Journal, 3, no. 2, pp. 136-143. (Reproduced in Siewiorek, Bell and Newell, 1982, pp. 711-715.)

Strecker, W. D. (1978). VAX-11/780 - A Virtual Address Extension to the DEC PDP-11 Family, AFIPS Proc. NCC, pp. 967-980. (Reproduced in Siewiorek, Bell and Newell, 1982, pp. 716-729.)

Tamir, Y. (1983). Strategies for Managing the Register File in RISC, IEEE Transactions on Computers, C-32, no. 11, pp. 977-988.

Thornton, J. E. (1964). Parallel Operation in the Control Data 6600, AFIPS Proc. FJCC, part 2, 26, pp. 33-40. (Reproduced in Siewiorek, Bell and Newell, 1982, pp. 730-736.)

von Neumann, J. (1945). First Draft of a Report on the EDVAC, Moore School, University of Pennsylvania.

Wilkes, M. V. and Stringer, J. B. (1953). "Microprogramming and the Design of the Control Circuits in an Electronic Digital Computer," Proceedings of the Cambridge Philosophical Society, part 2, 49, pp. 230-238. Cambridge University Press. (Reproduced in Siewiorek, Bell and Newell, 1982, pp 158-163.)

2
Research into Reduced Instruction Sets

During the 1970's most general-purpose computer families evolved towards configurations with common sets of capabilities. Data and address bus widths became 32 bits, virtual memory became standard, as did multi-tasking controlled by a protected operating system. Product differentiation centred on price and performance issues, with different technologies and cache organizations offering the different trade-off points.

Also during the 1970's, high level languages became the dominant programming method for minicomputers and mainframes. Computer architects added increasingly complex instructions into the CPU microprograms in order to minimize the semantic gap between the instruction set and the high level language.

As the microprocessor developed over the same period, it trailed the minicomputer and the mainframe in bus width and complexity. But semiconductor processing technology advanced relentlessly, and with each step forward the microprocessor designers could simply copy another stage of minicomputer development onto silicon. Around 1980, technology allowed a full 32-bit CPU to be built on one chip, and the minicomputer complex instruction sets were faithfully imitated by the microprocessor designers.

Then, from the west coast of America, this process was questioned. Is an optimal architecture for a VLSI processor necessarily going to be the same as an optimal architecture for a processor built from many chips on several printed circuit boards? Do these complex instruction sets actually give the best performance?

In this chapter we shall look at the research processor architectures which were built to answer these questions, and the alternatives which were proposed to the complex instruction sets of the minicomputers. These alternatives are now known as *Reduced Instruction Set Computers*, a name adopted from the work at Berkeley University.

2.1 THE IBM 801

In October 1975, a research group at the IBM Thomas J. Watson Research Centre began the design of a minicomputer system, including the hardware, the operating system, and a compiler. The project was named after building 801 which the team occupied. The aim was to achieve significantly better price/performance for high level language programs than was available from existing machines. The 801 was implemented in ECL logic, and was not a VLSI CPU. It was, however, instrumental in initiating the thought processes which led to the VLSI work at Berkeley and Stanford, and is included here for that reason.

2.1.1 The 801 Design Philosophy

The most radical departure from contemporary architecture taken by the 801 team was to accept the constraint that instructions should take only one machine cycle. The architects had observed that high level language compilers generated code comprised mainly of simple instructions (LOAD, STORE, BRANCH, COMPARE, ADD), and the performance of these was therefore critical to the system performance. If adding a complex instruction slowed down the CPU cycle, making all these simple instructions slower, that complex instruction had better add sufficient performance to overcome this loss and then show a benefit. In general, adding logic to allow the addition of a complex instruction cannot be justified against this criterion.

Usually minicomputers with complex instruction sets use microcode to allow the complex instructions to be executed on a relatively simple datapath in several sequential steps. Here the advantage gained by microcode over a subroutine of simple instructions is that the microcode is held in a memory with a very short access time compared with main memory. If the CPU has a cache memory, however, this advantage is lost. Microcode gives good performance to a pre-selected set of sequences of simple instructions, whereas a cache will give the same performance to a dynamically self-adjusting set of sequences.

Another advantage to having complex functions built out of simple instructions is that an optimizing compiler can potentially improve the sequences of instructions, whereas it clearly could not change a fixed microcoded sequence. It can also make better use of compile-time information to simplify special cases of complex functions. Furthermore, the removal of complex instructions ensures good interrupt response without recourse to interruptable and restartable instructions.

It is clear, however, that performing a LOAD in a single cycle will not generally be possible if the result is to be fetched to a register before the instruction completes. Therefore the LOAD instruction on the 801 merely issues the address and locks the target register during its execution cycle, and then the processor moves on to the next instruction. At some point in the future the result arrives to be placed in the target, and the register may be used thereafter. The 801 used this *delayed load* structure, and similar techniques for STORE and BRANCH, where the action of the instruction happens after the instruction has completed and a subsequent one has started.

The philosophy of the 801 team was to allow user access to the machine only through the compiler. By careful compile time, link time, and load time checking, combined with run time checking where necessary, it was possible to dispense with hardware protection mechanisms (such as supervisor state, multiple virtual address spaces, and memory protection hierarchies). System calls were implemented as standard high level language procedure calls, and most of the system was written in the same language (PL.8, a derivative of PL/I). The only assembly language programs comprised less than a thousand lines of supervisor code and some of the complex library functions.

2.1.2 Instruction Set

The data types supported by the 801 are characters (bytes), half-words (half-word aligned), and 32-bit words (word aligned). All instructions are one word. The instructions use a three address format, so that the destination register is specified independently of the two source registers. This allows both operands to be preserved for future use; if one of them is not needed again the compiler can specify the same register for the result, but if they are both needed then a copy instruction has been avoided. Each register specifier is 5 bits, and can therefore address any of the 32 general-purpose registers.

Arithmetic instructions may use a 16-bit sign extended immediate constant, and logical instructions may use a 16-bit logical constant as one of the operands.

LOAD and STORE instructions use Base + Index or Base + 16-bit sign extended Displacement addressing. The address may optionally be copied back into the base register for auto-increment and auto-decrement addressing. Bytes, half-words and words may be transferred, and half-words may optionally be sign extended.

BRANCHes may have their destination specified as a 26-bit absolute address, a 16- or 26-bit relative offset from the current program counter, or as the sum of two general-purpose registers. The current value of the program counter may be saved in a Link register for subroutine return purposes. There are explicit instructions for delayed and non-delayed branches.

Shifts may be of any number of bits up to 31, and field extraction and merging are supported.

The instruction set also includes Test under Mask and Conditional Branch instructions, as well as Compare and Trap, Multiply and Divide Steps, and primitives to improve the efficiency of decimal arithmetic and MAX and MIN operations.

2.1.3 Organization

The hardware organization of the IBM 801 is shown in figure 30. The processor is based around a bank of 32 32-bit registers. This bank has an unusually high number of ports: two for writing values into the bank, and three for reading values out. The three read ports allow simultaneous access to base, index and data registers for a STORE operation, and the two write ports allow a delayed LOAD data value to be written into the register bank independently of other datapath activity.

The machine has a two level pipeline. The first stage includes instruction decode, operand access, and ALU operation. The data for a STORE is also accessed during this phase. The ALU result is then latched or used as a LOAD or STORE address. The second pipeline stage includes shifting the ALU result from the previous cycle, setting the condition code bits, and returning the shifted result to the register bank. A LOADed value is also returned to the register bank during this phase, and forwarded directly to the ALU if needed immediately for the next operation.

All LOADs and STOREs transfer data from and to a data cache, which has a 32 byte line length. A separate instruction cache avoids the contention which would arise if a single mixed cache were used on a CPU such as this which requires a new instruction every cycle. This results in a consistency problem whenever self-modifying code is used, but high level language compilers do not use self-modifying code so long as satisfactory addressing modes are available, as is the case here. Self-modifying code is not totally avoidable, since loading a new program from disk is an act of code self-modification, but the caches are sufficiently exposed to software to allow explicit cache synchronization where necessary.

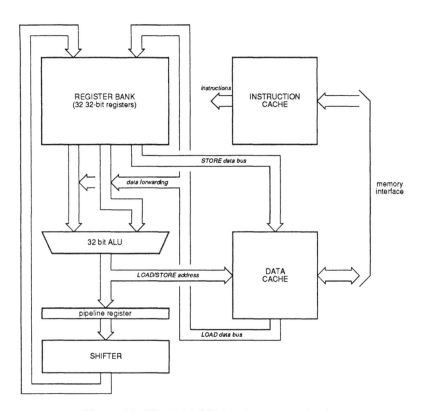

Figure 30: The IBM 801 hardware organization

2.1.4 Historical Perspective

The IBM 801 falls outside the scope of this book, as it is not a VLSI processor. It has been included primarily because of its historical place in the development of the RISC philosophy. Looking through Radin's (1983) description of the project one can see many of the important features incorporated which were to become crystallized in the subsequent academic projects at Berkeley and Stanford.

(1)   A large, multi-ported register bank with a LOAD/STORE architecture.

The 32 registers on the 801 do not seem excessive now, but at the time this was twice the normal complement. The register windowing scheme was introduced by Berkeley, and allowed the addition of arbitrarily large register sets. The 801 did not have this feature, and all 32 registers were uniformly addressable at all times.

The five ports on the 801 have not been generally copied; three or four is more usual. If, however, it is accepted that the most usual STORE instruction requires a base, an index, and the data to be stored, then three read ports are needed. With only two read ports, either an extra cycle must be used in the instruction to allow the three values to be accessed, or the addressing must be simplified to require only one register. The latter case then demands another instruction and another cycle to perform the full address calculation. Likewise for a LOAD, either a second port or a second cycle must be used to allow the value to be moved into the target register, if there is to be no conflict with a subsequent datapath operation which also writes a result to the register bank.

(2)   Dependence on advanced compiler technology.

RISC processors present special problems to compiler writers, especially when highly optimized code is required, but this does not mean that they are necessarily harder target architectures to compile onto. Many features of RISC CPUs actually ease the compiler writer's job. For instance, the fixed length instructions and the single branch mechanism both reduce complexity in selecting the right instruction to use for a particular job. On a RISC the choice of instruction sequence is often simplified by only having one option, or because all options are identical in code size and execution speed.

However, there are features of RISC designs that impose new disciplines. The LOAD/STORE architecture makes register usage critically important, and RISC CPUs have forced the advance of techniques in this area to minimize transfers to and from store. The combination of new compiler techniques and register windowing mechanisms has changed the proportion of data to instruction traffic from roughly 50:50 on conventional CISC architectures to 10:90 on some RISC architectures (though the RISC code expansion has contributed to increasing the instruction proportion also). Such a radical change in the nature of the memory traffic must also feed back into decisions on cache organization.

The other common RISC features introduced on the 801 which affect compilers are the delayed LOAD, STORE and BRANCH instructions, where the compiler must try to locate a useful instruction which can be moved into the delay slot.

The 801's dependence on the high level language for system protection has not gained general acceptance by other designers. Most commercial RISC CPUs have a protected supervisor state, and support multiple virtual address spaces. This may be because commercial systems must support a wide range of high level languages, and while a research team may trust one compiler, a commercial company is unlikely to be able to guarantee the security of a range of compilers. A compiler can in any case only protect against accident, and commercial systems must also offer protection against malice.

(3)   Pipelined single cycle execution of a simple instruction set, including delayed LOAD, STORE and BRANCH instructions.

The performance of a RISC comes mainly from breaking the execution unit into a pipeline of simple operations, and then keeping the pipeline full. Finding an instruction set which is a good match for the requirements of compilers at one end, and yet maps onto simple pipelined hardware at the other, is the key. The 801 fixed the instruction set to allow single cycle operation, yet still allowed base and index addressing. This success has not been matched by many VLSI processors since!

(4)   Incorporation of cache memories.

The 801 was built from a very fast technology (ECL), and was therefore ahead in cycle time of early VLSI RISC processors. The first integrated CPUs were coincidentally quite well matched to semiconductor memory speeds, and were therefore built without caches. This can now be seen as a temporary aberration. New RISC CPUs are built on 2 micron or smaller CMOS processes, and are substantially outstripping the speed of bulk memory devices. Progress over the next few years is probably going to be determined more by cache efficiency than the details of the CPU itself.

The 801 project pointed the way which was to be followed and expounded by the Berkeley and Stanford teams five years later. It also led to a VLSI implementation of a RISC architecture, the ROMP chip in the IBM 6150 RT PC. It is interesting to compare the 801 with the ROMP; in many ways the 801 is much closer to the archetypal RISC than its VLSI successor.

2.2 THE BERKELEY RISC ARCHITECTURE

The RISC project at Berkeley started from the observation that high level language compilers had difficulty making good use of complex instruction sets. This led to graduate studies of the behaviour of compiled programs to produce statistics about instruction frequencies, the use of addressing modes and local scalars, procedure nesting depth, etc. These statistics were used to guide architectural decisions in the design of a new instruction set suitable for VLSI implementation, and then the instruction set was implemented (twice) in VLSI by students on postgraduate courses.

The Berkeley RISC processors are characterized mainly by a very large register bank arranged in a novel window structure, whereby only a subset of the registers is visible at any one time. This organization is justified by its contribution to the minimization of memory data traffic during procedure entry and return. It should not be overlooked, however, that while adding this feature, the designers also took away a lot of complexity compared with a conventional (CISC) microprocessor. It is this adding of simplicity which has gained full acceptance amongst the RISC community; the windowed register file has found favour only with some of the other RISC developers.

Perhaps the most important contribution from the Berkeley team was their clear explanation and proclamation of their approach to the world at large. Their underlying research was thorough, their arguments convincing, and their VLSI designs were successful. Were it not for the fact that they appeared to be accusing the rest of the world (and large commercial concerns in particular) of getting their basic approach wrong, the RISC philosophy might have gained immediate acceptance, caused a slight change in the direction of computer development, and been forgotten. Their confrontational approach to the presentation of their conclusions may have delayed their widespread acceptance, but the effect on the industry has been correspondingly greater.

The example of the Berkeley work was undoubtedly a major contributory factor to the subsequent widespread development of RISC processors. Indeed, the term *Reduced Instruction Set Computer* was first used at Berkeley. Therefore we shall look at their arguments as well as their architecture. The implementation of the architecture that is presented here is that of their second VLSI chip, RISC II. This is described in detail by Katevenis (1985).

2.2.1 The Background to the Berkeley RISC Project

The Berkeley RISC processors were built to prove a point. The designers wished to demonstrate that developments in computer architecture were going down the wrong route, and especially so where the processor was constructed as a VLSI component. The silicon area on a chip is a limited and precious resource, and complex microcoded processors squander this resource on features which are little used in typical applications.

It is always hard to justify beyond question any detailed feature of a general-purpose processor architecture, especially in the design phase. This is because the utility of a feature can only be judged on the basis of its benefit in typical compiled high level language code. The definition of a typical application for a general-purpose computer is a source of some difficulty. In addition, if a compiler exists at all during the design phase, it is likely to be unrefined compared with later versions. Who is to say that a more advanced compiler cannot make better use of the feature in question?

The Berkeley argument addressed this issue very thoroughly, and space allows only a sample of their work to be discussed here. They studied computer design on the basis of two issues:

(1) *Function.* What sort of computations must it perform, and what features are necessary to make those computations efficient?

(2) *Cost.* Is it possible to implement the desired features in a cost effective manner in a particular technology, and what compromises are required to do so?

A *general-purpose* computer was assumed to be one that was required to perform a range of operations without the design being optimized for any particular one. Word processing, data base applications, mail and communications, compilations, computer aided design, control and numerical applications were expected to be represented in a typical mix of activities.

Information from previous studies of instruction set usage was collated with new results to paint an overall picture of what computers spend most of their time doing. The conclusion was that the general trend towards increasingly complex instruction sets was misdirected. The case was stated most strongly by Patterson and Ditzel (1980), and was based on the following observations regarding complex instruction sets:

(1) *Complex instructions are used infrequently by compilers.* A particular IBM 360 compiler produced code where 10 instructions accounted for 80% of all instructions executed, 16 for 90%, 21 for 95% and 30 for

99% (Alexander and Wortman, 1975).

(2) *They result in irrational performance characteristics.* The INDEX instruction on the VAX-11/780 can be replaced by several simple instructions with a 45% speed improvement. If the compiler knows that the lower bound is zero, the improvement can be 60%.

(3) *Complex machines take longer to design.* The designers must therefore predict further into the future and pioneer the target technology, or the product will be launched with a relatively old technology.

(4) *They are likely to contain more design errors.* Mainframe computers now generally use writable microcode stores, and bug fixes are issued like fixes to operating system code. A bug in the microcode ROM on a VLSI processor is harder to deal with.

(5) *Complex instructions slow down the whole machine.* The propagation delay of a VLSI logic gate has a strong dependency on the load being driven. Larger fan-outs mean slower logic. Adding a complex instruction will typically mean increasing the average gate fan-out and slowing the machine cycle. Large control PLAs are slower than small ones. If a complex instruction slows the machine down by 10%, it must increase the average throughput per cycle by more than 10% to be cost-effective. There is little evidence to support the inclusion of complex instructions on this basis.

(6) *There are better things to do with the chip area.* Even if the semiconductor technology enables the construction of a complex instruction set processor on a single chip, it may still be better to build a simple processor and use the area saved for larger (and faster) transistors, pipelining, or on-chip caches.

The conclusion is that VLSI processor design should start from a minimal simple instruction set. Instructions should be added only if they can be justified on the basis that either they are cost-effective, or they are necessary and cannot be synthesized from the simple instructions (for example Supervisor Call). Any chip area left over should be used for performance enhancing features such as cache memory.

The Argument for the Windowed Register File

One set of program analyses (reported by Patterson and Sequin, 1981) focused on four C programs and four programs written in Pascal. These showed that typically 60% of the data types referenced were scalar variables, with the remainder being divided between integer constants (18%) and arrays or other structures (22%). Over 90% of the scalars were local values, whereas over 80% of the arrays and structures were global. The occurrence rates of various high level language constructs were also measured, and the results are summarized in table 3.

Table 3. Weighted relative frequency of HLL statements

measure:	occurrence		weighted by # instrs.		weighted by # mem. refs.	
HLL:	Pascal	C	Pascal	C	Pascal	C
call/return	12	12	30	33	43	45
loops	4	3	40	32	32	26
assign	36	38	12	13	14	15
if	24	43	11	21	7	13
begin	20	–	5	–	2	–
with	4	–	1	–	1	–
case	1	1	1	1	1	1
goto	–	3	–	0	–	0

(Source: Patterson and Sequin, 1981. ©1981 IEEE.)

The numbers of instructions and memory references for each construct were based on information from compilers for the VAX, PDP-11 and Motorola 68000. The CALL statement included parameter passing and saving and restoring general registers and the program counter. The IF and CASE statements included instructions to evaluate the associated conditions, and to jump. The loop statements included all the instructions executed during each iteration.

In a typical memory bandwidth limited system, the number of memory references is the best indicator of the time cost of a statement. It is therefore very significant that procedure call/return is promoted to first place in this table when weighted according to the memory reference criterion. Other studies provided further evidence that procedure call overheads are very significant in typical high level language programs.

Therefore a processor architecture for high level language programs should be optimized for procedure call and return, and it should offer efficient handling mechanisms for local scalar quantities. The Berkeley RISC architecture starts from this observation. The large windowed register file fulfils both of these requirements.

2.2.2 The Berkeley RISC II Programmer's Model

The register organization of RISC II is shown in figure 31. A very large bank of registers (138 32-bit registers in all) is divided into overlapping windows, so that at any time exactly 32 of them are visible to the user. The visible subset is determined by the *current window pointer* (CWP) in the program status word. On procedure entry CWP is decremented, and on return it is incremented. The registers in the overlap region are used to pass parameters from one procedure to the next, and the moving of the window exposes new registers for use by the new procedure, so that the overhead of saving workspace registers on each procedure entry is avoided.

In fact, not all of the visible registers are affected by CWP. Registers 0 to 9 are *global* registers, which contain values that are required at all procedure levels. (Register 0 always contains the value zero, and writing to it has no effect. This fact is used to avoid the need for explicit compare instructions, as a subtract to destination register 0 has exactly the desired properties.)

Other registers which are visible to the user include the *next PC*, which is the address of the instruction currently being fetched, and the PC itself, which is the address of the current instruction. The *last PC* is available only to programs which run in supervisor mode, and is used for saving state on exception entry.

The final bit of user state is contained in the 13 bit program status word (figure 32). This contains the conventional ALU flags for zero result, negative result, arithmetic overflow and output carry. The interrupt enable and supervisor mode bits are also conventional. The current window pointer bits are used to determine the active register window as described above. The previous system mode bit is used to retain a copy of the S bit upon exception entry. The *saved window pointer* is used to cause a trap whenever CWP

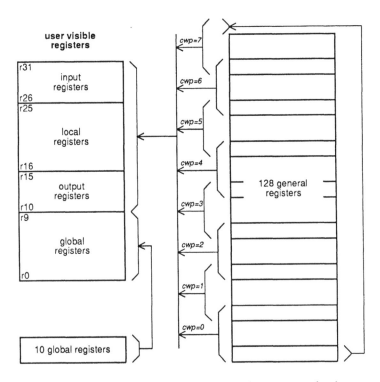

Figure 31: Berkeley RISC II register window organization

increments or decrements to the value of SWP, thereby alerting the system software to the fact that more windows need to be saved or restored.

2.2.3 The Berkeley RISC Instruction Set

The instruction set uses two basic formats (figure 33), with a 13-bit immediate field for instructions with explicit source register addressing and a 19-bit immediate field for instructions with implicit (PC relative) addressing. Conditional instructions (jumps and returns) use an implicit destination, and encode the condition in the field normally used for the destination register address. The short (13-bit) immediate form may specify a third register in place of the immediate value, and this is allowed as the second ALU operand or as the shift amount specifier. Note that there are several unused bits in the three register form of the instructions.

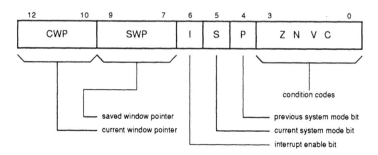

Figure 32: Berkeley RISC II program status word
(From Katevenis, 1985)

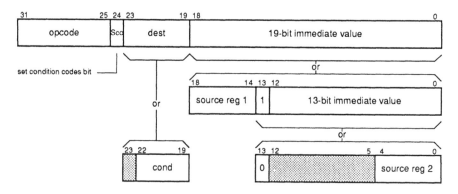

Figure 33: Berkeley RISC instruction formats
(From Katevenis, 1985)

All instructions have an SCC bit which determines whether or not the condition codes should be modified by the instruction. The setting of the codes is conventional for the arithmetic and logical instructions, and can be used to detect zero and negative data values during a load. It is generally meaningless at other times, and should be disabled.

The instruction set is listed in table 4.

RISC II uses a delayed jump mechanism, so that the instruction which is fetched during the execution phase of the jump is executed whether or not the jump is taken.

The CALL instructions perform an unconditional jump, and in addition they copy the PC into the destination register and decrement CWP. The return instructions only support the register relative jump, and they increment CWP. Modifying CWP either way can cause a window over- or underflow exception.

2.2.4 Operation of the RISC II Register Windows

In normal operation, each register window contains a procedure activation record. The overlapping region between two windows contains the output parameters of the parent procedure, which are also the input parameters to the child procedure. The registers therefore contain the top of the current stack.

All the arithmetic on CWP is performed modulo 8, and window 7 overlaps window 0, so the windows operate as a circular buffer. Obviously, CWP cannot be allowed to go round and round the buffer unchecked, since beyond a nesting depth of 8 procedures the record for the top procedure level will be overwritten and lost.

The solution lies with SWP, the Saved Window Pointer. This points to the last window which has been saved onto a stack in main memory. If a CALL instruction causes CWP to decrement to the same value as SWP, so that the process is about to use the last window, an exception is generated. The trap handler can then save the window below SWP to the external stack, decrement SWP, and resume the CALL. Similarly if a RETurn instruction causes CWP to increment to equal SWP, the saved window can be reloaded from the external stack, SWP incremented, and the process resumed. System software can give the user the impression of having an unlimited supply of register windows.

If every call and return traps to save a window then the window mechanism has no benefit. This would be the case if, for instance, procedure nesting depths displayed rapid variations over large ranges. Calculating factorials by the obvious recursive algorithm would be a bad case in point. Typical programs do not display such behaviour, however. Analyses (summarized in Katevenis, 1985) suggest that with 8 windows available, 1 to 3% of procedure calls will cause a window overflow. Smaller numbers of windows

Table 4. RISC II instruction set

Arithmetic and logical operations

(Three registers, or register + 13-bit signed immediate to register.)

AND,OR,XOR
ADD,SUB   *(with or without carry in)*
SUBI   *(reverse subtract, with or without carry in)*
S   *(shift left logical, right arithmetic or logical)*

Transport operations

(Load or store bytes, half-words or words.
 Loads optionally sign extend bytes and half-words.)

LDX   *(load at base + 13-bit signed offset or base + index)*
LDR   *(load at PC + 19-bit signed offset or PC + index)*
STX   *(store at base + 13-bit signed offset)*
STR   *(store at PC + 19-bit signed offset)*

Control transfer

JMPX   *(conditional jump to register + 13-bit signed offset or index)*
JMPR   *(conditional jump to PC + 19-bit signed offset)*
CALLX   *(call to register + 13-bit signed offset or index)*
CALLR   *(call to PC + 19-bit signed offset)*
RET   *(conditional return to register + 13-bit signed offset or index)*
RETI   *(conditional return from interrupt; privileged use only)*
CALLI   *(call interrupt routine; hardware use only)*

(continued)

Table 4. (continued)

Special instructions

LDHI	*(load 19-bit immediate value into top bits of register)*
GETLPC	*(save last PC into a register; privileged use only)*
GETPSW	*(copy PSW)*
PUTPSW	*(update PSW; privileged use only)*

(Source: Patterson and Sequin, 1981. ©1981 IEEE)

perform worse; with 4 windows available, overflow might happen on up to 15% of the procedure calls. Below 4 windows the overhead of handling the exception would probably be so high that a single window with conventional stacking would be more efficient.

It is possible for the window saving routine to save more than one window when an overflow occurs, so that the next nested call will not trap. Tamir and Sequin (1983) published a study of the effectiveness of various algorithms, and concluded that for up to 8 register windows saving a single window is the best overall strategy. Larger window sets might benefit from more windows being saved on each trap.

The exception algorithm described above causes a trap whenever CWP increments or decrements to equal SWP. This appears to leave the window addressed by SWP unused, which is wasteful. In fact one window must be left free for exception handling (including window overflow), so this system is optimal, and requires a single comparison to generate both trap conditions. If the interrupt handler is to be re-entrant, it must ensure that the next window is saved before enabling a subsequent trap.

2.2.5 The Implementation of RISC II

The datapath organization of RISC II is shown in figure 34. The various major components are described briefly below:

- The *register file* is very large, so the design of the basic one bit cell is very important. In particular, the number of buses running through the

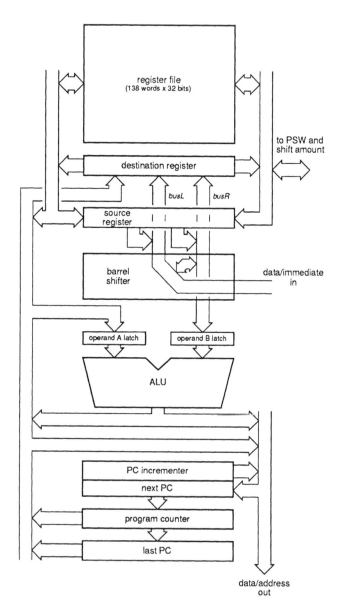

Figure 34: RISC II datapath organization
(From Katevenis, 1985)

cell has a great effect of the area of the cell. The RISC I chip had a conventional three bus register cell. RISC II used a two bus cell, where reads used a single bus (so two operands could be fetched simultaneously), and writes used both buses together. The register cell used a slightly modified six transistor static RAM cell, with single-bus read sensing (whereas static RAMs normally use differential sense amplifiers across both buses).

The two bus register cell on RISC II was 2.5 times smaller than the three bus cell used on RISC I, and this was the main reason why the second implementation of the architecture was developed.

- The *destination register* is a temporary pipeline latch where information is stored prior to being written into a register or used elsewhere.

- The *source register* holds the input to the barrel shifter.

- The *barrel shifter* was built with pass transistors which allow signals to pass in either direction, and this was exploited on RISC II. Data flows in through *busR* (which passes straight across the matrix) and out through *busL* (which runs diagonally across the matrix) for right-to-left shifts, and in the reverse direction for left-to-right shifts. The shifted value is latched in the destination register or in the operand B latch.

Immediate values are introduced onto the datapath down *busL* where it emerges at the side of the datapath after crossing the matrix.

- The *ALU* has two input latches, and performs ADD, SUB, AND, OR, XOR, or transmits the B operand to the output. The result is held in the destination register or stored dynamically on the data/address output bus.

- The *program counter* contains the address of the instruction which is currently executing on the datapath. The next PC is the address of the instruction currently being fetched from memory, and the last PC is the address of the instruction last executed. If an exception arises, last PC will hold the address of the aborted instruction during the first cycle of the exception sequence.

A dedicated incrementer computes the value of next PC plus two or four bytes, for normal sequential instruction fetching (the two byte increment is used in conjunction with Expanding Instruction Caches; see Patterson et al, 1983). Branch and load/store addresses are computed in the ALU.

Pipeline Operation

The operation of the RISC II pipeline is illustrated in figure 35. The usage of the various datapath facilities by the first instruction (I1) is shown by shading.

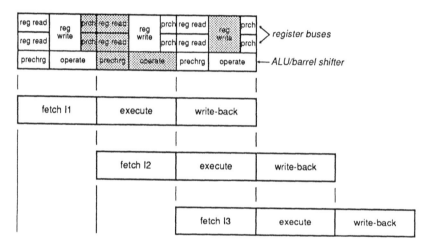

Figure 35: RISC II pipeline operation
(From Katevenis, 1985)

The chip uses a three stage fetch/execute/write-back pipeline, with a new instruction starting every cycle. The two bus register file presents an interesting difficulty, since in any cycle the buses may be needed for operand fetching by one instruction, and for writing a result back to the register file by the preceding instruction.

The problem is resolved by dividing the cycle into three time slots, allocated to reading, writing, and precharging the buses. Every read must be preceded by a precharge operation, since the bus is driven single ended by a register cell. The register cell must have a relatively weak drive capability if it is to have small size. The write operation does not require precharge, however, as it uses both buses driven differentially by the result register. The result register can have large transistors to drive the buses as this cost is only incurred once per data bit; a similar cost in the register cell would be incurred 138 times per data bit!

Therefore the cyclic sequence of register bus activity must be precharge, then read, then write. To allow a reasonable time for the ALU to produce its result, the execute cycle cannot use the write phase of the same cycle; it must defer the result in a pipeline register and use the write phase of the following cycle.

This is all well and good unless the next cycle tries to use the result of the current cycle as an operand. In this case register forwarding logic detects the clash of register usage and passes the result back to the ALU (or shifter) for re-use before writing it back to the destination register.

The Datapath Control Logic

The logic which controls the operation of the datapath is illustrated in figure 36. The direct use of the instruction fields is evident here; in many cases the control logic consists solely of a set of pipeline latches which delay the instruction bits until the appropriate cycle. The register forwarding logic just compares the destination register at stage 3 with the source registers at stage 2 (which holds values for the instruction following that at stage 3), and enables the datapath forwarding route when a match is found.

The latches which provide the addresses to the register decoders must operate as stage 3 latches for the destination register, and as stage two latches for the operand registers, since the same decoders are used together for the former and individually for the latter operation.

2.2.6 A VLSI Cache for RISC II

A separate VLSI instruction cache was developed by a team on a postgraduate architecture course, and this component embodied a number of interesting features (Patterson et al, 1983):

(1)   It had a *remote program counter*. The cache contained an estimated copy of the program counter, and used it to initiate cache accesses before the real PC was available. By the time the cache data and tag were ready, the real PC was valid, and could therefore be used for the tag comparison.

Sequential instruction execution is easy to predict, but it does not continue indefinitely. The cache chip could identify jump instructions and move automatically to the target. Conditional branches present a problem for this sort of mechanism, but on RISC II such instructions

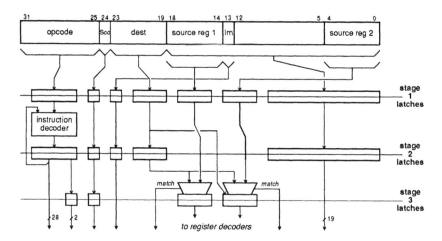

Figure 36: RISC II control structure
(From Katevenis, 1985)

contain an unused bit which can be used by a compiler to give a hint to the cache whether or not the branch is likely to be taken.

(2)   The cache used *compacted instructions*. The RISC II instruction set gives code of relatively low density, because it has been optimized for very easy decoding in hardware. The density can be improved by using special short encodings for frequently used instructions, but these are then harder to decode. If, however, the decoding is performed when the instruction is fetched into the cache, this overhead occurs once however many times the instruction is subsequently referenced (so long as it remains in the cache).

The proposal here was to exploit the many unused RISC II opcodes in order to define a set of 16 bit instructions which would be expanded into the corresponding 32 bit instructions when they were fetched from memory. The processor sees the simple format, but the memory size and bandwidth requirements are set by a denser code.

(3)   The cache was *expansible*. Multiple cache chips could be attached to one RISC II processor. A simple direct mapping decoder selected a particular chip for any access.

(4)   It was *fault tolerant*. A second valid bit was associated with each cache line. This bit was loaded at system initialization, and any defective line was marked as permanently invalid. An access to that line would always miss, causing a reference to main memory. Though this approach has a performance penalty associated with it, it does allow chips with certain minor defects to be used whereas they would otherwise be rejected.

Though this cache chip was not central to the RISC developments at Berkeley, it further illustrates the wealth of ideas which were being generated by the students and staff in the early 1980's. The above features will all recur in various guises in the next chapter when we look at the commercial exploitation of the RISC idea.

2.3 STANFORD MIPS

At about the same time that the RISC development was proceeding at Berkeley, a parallel development was taking place at Stanford University. Their common theme was the goal of speed through simplicity, but the approaches were very different.

The Stanford processor was called MIPS, which stood for *Microprocessor without Interlocked Pipeline Stages*. The philosophy underlying the design was to move complexity from hardware to software wherever it was practical to do so, and as the name implies, this philosophy was applied to the management of the data dependencies in the processor pipeline. The hardware had no interlocks to control data race hazards; these were left entirely as a problem for the compiler support software to handle.

The machine level instruction set of MIPS is also unusual. Each instruction is 32 bits long, and may contain the functionality of two completely unrelated user level instructions. Because of the complexities of the dependencies which have to be managed at the machine instruction level, a more normal abstract instruction set is defined for programming the processor, and a program called the *Reorganizer* converts the abstract (user) instruction set into machine instructions. All the dependencies are handled by the Reorganizer.

2.3.1 The MIPS Instruction Set

MIPS does not have a large register file like the Berkeley RISC processors; it has a more conventional bank of 16 general-purpose 32-bit registers. In addition, special registers are used to support multiply and divide step instructions, past, present, and future PC values, and address masking for memory management.

The processor uses word addressing only. Byte operations are performed internally with byte insert and extract instructions, but the external memory need only support word transfers. There is logical support for 32 bit virtual addresses, but the chip has only 24 address pins, so the memory space is restricted to 64 Mbytes.

The instruction set is defined at two levels. The user level instruction set uses a fixed 32-bit format, and has no pipeline dependencies or delayed branches. The instructions are listed in table 5.

The machine level instruction set (which is generated only by the Reorganizer) splits the user instructions into *pieces* of various lengths, and

Table 5. MIPS user instruction set

Arithmetic and logical operations
*(both two and three register forms are used;
the second operand may be an 8-bit immediate value)*

ADD,SUB,AND,OR,XOR

SUBR	*(reverse subtract)*
IC,XC	*(byte insert and extract)*
RLC	*(rotate combined register pair and extract 32 bit field)*
ROL	*(rotate left)*
S	*(shift left logical, right arithmetic or logical)*
MSETUP	*(multiply set up)*
MSTEP	*(2-bit multiply step)*
UMEND	*(unsigned multiply end)*
DSTEP	*(divide step)*
SET	*(set register to 0 or -1 depending on condition test)*

Transport operations

LD	*(load immediate, absolute, base + 20- or 8-bit offset, base + index, scaled base)*
ST	*(store absolute, base + 20- or 8-bit offset, base + index, scaled base)*
MOV	*(immediate byte or register)*

Control transfer

BRA	*(branch PC relative, optionally conditional)*
JMP	*(jump absolute, register relative, register relative indirect)*
TRAP	*(conditionally trap to location zero)*

(Source: Hennessy, 1984. ©1984 IEEE.)

then packs multiple pieces into one machine instruction word. The various pieces are categorized as follows:

- *ALU pieces.* These are the register to register instructions, with two or three operands. They typically require less than half an instruction word. They include byte insertion and extraction, a two bits per cycle Booths multiply step, and a one bit per cycle divide step. A two operand piece can be packed into one instruction with a three operand piece.

- *Load/Store pieces.* These require 16 to 32 bits of the instruction word, and the shorter ones can be packaged with an ALU piece.

- *Control Flow pieces.* All of these have a delay slot.

- *Special instruction pieces.* For procedure and interrupt linkage.

The machine instruction formats are shown in figure 37. Note how two ALU operations may be packed into one instruction, one using a three address form and one a two address form. The conditional branch instructions use the first ALU operation to add the offset to the PC, and the second to evaluate the condition. Full details of the machine level instruction set are in Gill et al, 1983.

31 28	27 24	23 20	19 16	15 12	11 8	7 4	3 0	
opcode	src/dst1/ sub-opcode	24-bit signed immediate/offset/direct address						LD/ST/ BR/JUMP
opcode	src/dst1/ sub-opcode	base	20-bit signed offset					LD/ST/ JUMP
opcode	sub-opcode	cond	srcB2	srcA2	12-bit signed offset			BR/TRAP COND
opcode	sub-opcode	cond	dst/srcB2	srcA2	000000000000			SET
opcode	src/dst1/ sub-opcode	base	dst/srcB2	srcA2	alu op2	8 bit signed offset		LD/ST/JUMP +ALU
opcode	dest1	alu op1	dst/srcB2	srcA2	alu op2	srcB1	srcA1	ALU +ALU
opcode	src/dest1	base	dst/srcB2	srcA2	alu op2	sub-opcode	index/ shift	LD/ST +ALU

Figure 37: MIPS instruction formats
(From Gill et al, 1983)

The processor includes support for page faults, interrupts, and traps generated by arithmetic overflow or software interrupt. The old PC registers must be saved on exception entry to allow the program to be resumed later.

MIPS does not have a condition code register; instead it has single instruction compare and branch operations, and conditional register set or clear instructions.

The chip produces virtual addresses, and a virtual memory system requires an off-chip translation mechanism. There is on-chip support for multiple processes, however. This support is in the form of an address masking unit (figure 38), which replaces some of the top bits of the virtual address with a process identifier. The replaced bits must be all zeroes or all ones, otherwise an exception is generated. A process can therefore grow from the top and the bottom of its virtual address space. If it requires more memory than was initially allocated to it, it can be given a new process ID which uses fewer bits, thereby releasing more virtual address bits for addressing memory. This can be done as part of the address exception handling software.

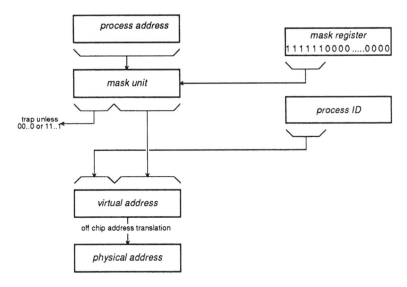

Figure 38: The address masking mechanism on MIPS
(From Hennessy, 1984. ©1984 IEEE.)

2.3.2 The Implementation of MIPS

A block diagram of the MIPS implementation is shown in figure 39. Two 32-bit busses connect all the functional units together. The ALU uses a full 32-bit carry look-ahead scheme; it can add two 32-bit numbers in 80 nanoseconds. The barrel shifter supports byte insertion and extraction in addition to multi-bit shifts and rotates, and it will extract a 32-bit field from any bit position in the 64-bit concatenation of two registers. It can't, however, be used at the same time as the ALU.

There is hardware support for multiply and divide step instructions in the form of two special registers which hold the intermediate results.

The old PC registers keep a trail of the three most recently used PC values, and must be saved on exception entry to allow the process to be restarted. If the exception happened between a branch and the instruction in the delay slot, for instance, the process is resumed by branching to the delay slot instruction. This branch will have its own delay slot, and that should contain a branch to the original branch destination.

The processor uses a 5-stage pipeline, with a new instruction starting every 2 cycles (figure 40). The stages are:

(1)   IF - instruction fetch and increment PC.

(2)   ID - instruction decode.

(3)   OD - operand decode (compute the effective address using the ALU, and send it to memory memory).

(4)   OS/EX - operand store to memory; execute the ALU operation.

(5)   OF - operand fetch from memory (if the instruction includes a load).

Note that both OD and OS/EX make use of the ALU, so each machine instruction has two cycles of ALU usage. If the instruction is a package of two ALU operations, this is when they are executed. If a load is packaged with an ALU piece, OD is used for the address calculation and OS/EX for the ALU piece. A condition test and branch uses the OD slot for calculating the branch target (which is then placed in the branch target register), and the OS/EX slot is used to evaluate the condition test. If the condition is true, the branch target is copied into the PC, otherwise it is ignored.

As has already been mentioned, there are no hardware interlocks in this processor. The software is expected to know when loads have happened and when registers are up to date. Considerable work went into developing the software infrastructure which is necessary to make such a scheme practical,

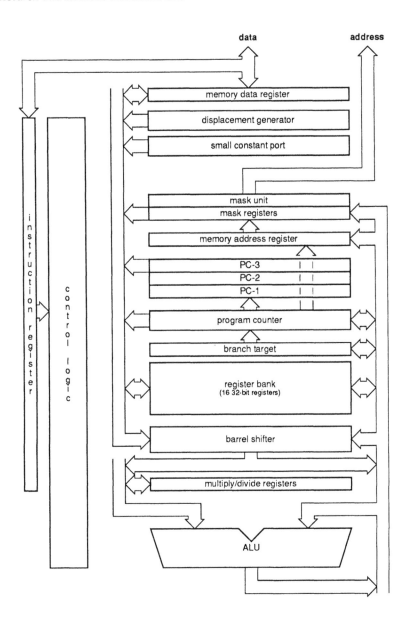

Figure 39: Block diagram of the Stanford MIPS processor
(From Hennessy, 1984. ©1984 IEEE.)

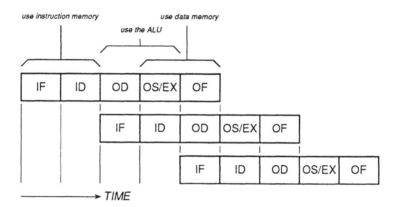

Figure 40: The MIPS pipeline operation
(Adapted from Hennessy et al, 1981)

and some of this is described in the next section.

The chip was built on a 4 micron NMOS process, and used 24,000 transistors in an 84 pin package. It achieved a 250 nanosecond cycle time, at which speed it delivers 2 MIPS. Separate external memories were used for instructions and data in order that the memory could keep up with the processor without a cache being required.

MIPS had no explicit support for floating-point operations, either in hardware or in software. Gross (1985) describes a software implementation of a set of routines for single precision floating-point arithmetic, using the standard integer instructions, and comments on the usefulness of several of the instructions (including one instruction which was ultimately left out of the implementation).

2.3.3 MIPS Software Issues

The MIPS architecture requires a code Reorganizer to reorder instructions to ensure that an instruction is never executed before its operands become available. This is a result of the quite deliberate decision to build the processor without the hardware interlocks which are conventionally used to solve the dependency problem. Moving the solution from hardware to software is justified on the basis that hardware interlocks always have a performance cost, which must be paid whenever the code is executed, whereas the software solution has a one-time cost at compile time.

The required code reordering problem may be complex; indeed, Hennessy and Gross (1983) show it to be a member of the NP-complete class of problems, which implies that finding a provably optimal solution will only be practical for trivially short sequences of instructions. They propose a heuristic algorithm to solve the problem. In general, however, producing a provably optimal instruction sequence for any processor is an intractable problem, and while the absence of interlocks obviously makes the code generation harder, it is clear that it does not move the problem into a different order of computational difficulty.

The simplest way to turn a sequence of user instructions into legal code is to add no-ops after every instruction which has a delayed effect, so that the processor idles until the instruction has completed. For instance, consider the user code sequence (from Hennessy and Gross, 1983):

```
1        Load  1(sp),R1
2        Add   #1,R1
3        Store R1,1(sp)
4        Load  2(sp),R2
5        Add   R1,R2,R3
6        Store R3,A
```

This is transformed into a legal sequence by the addition of No-Ops:

```
1        Load  1(sp),R1
2        No-Op
3        No-Op
4        Add   #1,R1
5        Store R1,1(sp)
6        Load  2(sp),R2
7        No-Op
8        No-Op
9        Add   R1,R2,R3
10       Store R3,A
```

The Reorganizer can then attempt to reorder the instructions to replace the No-Ops with useful instructions which do not depend on the result of the delayed instruction:

```
1        Load 1(sp),R1
2        Load 2(sp),R2
3        No-Op
4        Add  #1,R1
5        Store R1,1(sp)
6        Add  R1,R2,R3
7        Store R3,A
```

This example illustrates the code reordering required to cope with pipeline dependencies, and shows how the delay slots may often be used for useful work. It has been shown (Hennessy, 1984) that 21% of all MIPS instructions are executed in a branch delay cycle. The example does not illustrate the packaging of multiple pieces into a single instruction, though the MIPS Reorganizer does this as well.

The heuristic algorithms used to optimize the code reordering are based on the analysis of directed acyclic graphs, where each node is a particular user instruction and the arcs represent data dependencies. The details of the algorithms are complex, like many code optimization algorithms. There is the additional difficulty that it is harder to tell by inspection whether the output code is correct.

The final difficulty arises when the program is compiled, and a debugger is needed to locate an error in the source code. The correspondence between the object and source codes may be hidden by the reordering, rendering debugging difficult. This problem is not very different from that which arises with any optimized compiler output, however.

2.4 CONCLUSIONS

The Berkeley team rewrote the rule book of VLSI processor design. The simple instruction set and pipelined datapath operation have been adopted by many microprocessor design teams. The complex instruction set microprocessor is far from dead, but it has certainly lost the high ground in the battle for maximum performance.

The dominant feature of the Berkeley RISC architecture, the large windowed register file, has not gained such universal acceptance. Though it does reduce the data traffic caused by procedure entry and exit, it also increases the overhead of context switches, and occupies considerable silicon area. The Sun SPARC is a commercial RISC processor architecture which embodies this feature in a clearly recognizable form, and some Japanese CISC processors (eg the Hitachi H16) also have a similar register organization, but other RISC designers have chosen different organizations and sizes for their register sets. The register windows concept is a worthy architectural idea in its own right, but it is not related directly to the RISC versus CISC debate.

The *Reduced Instruction Set Computer* may have been born out of earlier ideas, but it came of age at Berkeley, and it is due to the research, persuasion and demonstration of the Berkeley team that it has grown into a major force in the world microprocessor market.

The Stanford MIPS processor stands alongside the Berkeley RISC processors as a vanguard of the RISC movement. The driving philosophy behind its design is the same as that at Berkeley, but the interpretation of that philosophy was different from that made at Berkeley. Some common factors emerged, however. Most notable amongst these were:

- A load/store architecture.

- Single cycle pipelined execution of most instructions. (MIPS executes a compound double instruction every two cycles.)

- Delayed branches.

These common factors can all be traced back to the IBM 801 project. To these Berkeley added their very large windowed register file, whereas Stanford moved towards increased hardware simplicity by moving interlocks into software. Neither of these features has gained full acceptance, but both have been taken up and used in commercial designs.

The original MIPS ideas have continued to be developed. The Stanford academics have developed a similar processor with a 2Kbyte on-chip instruction cache and around 20 MIPS performance (MIPS-X, see chapter 5). A commercial venture called MIPS Computer Systems, Inc., has also grown out of the Stanford work and produced a very high performance family of processors (eg the MIPS R2000, see chapter 3).

References

Alexander, W. C. and Wortman, D. B., (1975). Static and Dynamic characteristics of XPL programs. Computer, 8, no. 11, pp. 41-46.

Gill, J., Gross, T., Hennessy, J., Jouppi, N., Przybylski, S. and Rowen, C., (1983). Summary of MIPS Instructions, Stanford University Technical Note no. CSL-83-237.

Gross, T., (1985). "Floating-Point Arithmetic on a Reduced-Instruction-Set Processor," Proceedings of the 7th Symposium on Computer Arithmetic, pp. 86-92.

Hennessy, J. L., (1984). VLSI Processor Architecture, IEEE Transactions on Computers, C-33, no. 12, pp. 1221-1246.

Hennessy, J. L. and Gross, T., (1983). Postpass Code Optimization of Pipeline Constraints, ACM Transactions on Programming Languages and Systems, pp. 422-448.

Hennessy, J. L., Jouppi, N., Baskett, F. and Gill, J., (1981). "MIPS: A VLSI Processor Architecture," Proceedings of the CMU Conference on VLSI Systems and Computations, pp. 337-346. Computer Science Press, Rockville, MD. (Also a later version as Technical Report 83-233, Stanford University, June 1983).

Katevenis, M. G. H., (1985). Reduced Instruction Set Computer Architectures for VLSI, MIT Press, Cambridge, MA.

Patterson, D. A. and Ditzel, D. R., (1980). The Case for the Reduced Instruction Set Computer, Computer Architecture News, 8, no. 6, pp. 25-33.

Patterson, D., Garrison, P., Hill, M., Lioupis, D., Nyberg, C., Sippel, T. and VanDyke, K., (1983). "Architecture of a VLSI Instruction Cache," Proceedings of the 10th Annual Symposium on Computer Architecture, ACM SIGARCH CAN 11.3, pp. 108-116.

Patterson, D. and Sequin, C., (1981). "RISC I: A Reduced Instruction Set VLSI Computer," Proceedings of the 8th Annual Symposium on Computer Architecture, ACM SIGARCH CAN 9.3, pp. 443-457.

Radin, G., (1983). The 801 Minicomputer, IBM Journal of Research and Development, 27, no. 3, pp. 237-246.

Tamir, Y. and Sequin, C., (1983). Strategies for Managing the Register File in RISC, IEEE Transaction on Computers, C-32, no. 11, pp. 977-989.

3
Commercial VLSI RISC

In this chapter we shall look in detail at a number of processor architectures developed by commercial organizations (though they are not necessarily commercially available). In the last chapter we examined the research origins of RISC, and saw three machines with some common features, but considerable differences also. These differences underline one of the difficulties of studying RISC architecture - what defines whether or not a particular architecture qualifies as a RISC?

Ultimately, the answer to this question is of little interest. RISC and CISC processors perform the same function in a system, and the system designer will choose his CPU on the basis of cost, performance, and ability to support the system hardware and software requirements.

The classification of a processor architecture as a RISC or a CISC is probably of primary interest only to the author of a book such as this. Which architectures should be included, and which left out? The criterion used here has been fairly broad. An architecture has been accepted for inclusion if it has at least one VLSI implementation, and either the manufacturer considers it to be a RISC (or an 'improved' RISC), or the architectural approach has a common thread with the RISC approach. An additional criterion is that there should be sufficient information available for publication to allow a reasonable insight into how the chip works. The qualifiers are rich in architectural and organizational variety.

The processors are grouped in this chapter in loose categories. We start with the simpler CPUs, the IBM ROMP and the Acorn RISC Machine. Then we introduce the two whose ancestry is most clear, from Sun SPARC (descended from the work at Berkeley) and the MIPS R2000 (from the work at Stanford). The two machines which support the stack cache model are described next (the AT&T CRISP and the AMD Am29000), followed by the more complex processors (the HP precision architecture, the Motorola M88000 and the Intel 80960KB). Finally we look at two designs which have rather more tenuous connections with RISC ideas, the Intergraph Clipper and the Inmos T800 transputer.

3.1 THE IBM 6150 RT PC

The IBM 6150 RT PC uses a proprietary RISC CPU called the ROMP (Research Office products division MicroProcessor). This design developed alongside the 801 described in the last chapter, but the ROMP processor is a single VLSI chip whereas the 801 CPU was built from ECL. The ROMP design diverged somewhat from the 801 in order to match product requirements. For instance, the dependence of the 801 organization on fast caches was not compatible with the RT PC product cost targets. This led to a reassessment of many aspects of the design, and resulted in the ROMP being quite different from the 801 (see Radin, 1983 for details of the 801 and IBM, 1986 for details of the IBM 6150 RT PC and the ROMP chip)

The ROMP and MMU chips are built on a 2 micron double-metal NMOS process, with level sensitive scan-path logic to simplify testing. The ROMP uses 45,000 transistors, the MMU uses 62,000 transistors. With a 6MHz clock the system delivers around 2 MIPS sustained performance.

3.1.1 ROMP Architecture

The ROMP has sixteen general-purpose 32-bit registers. In addition, it has a set of system control registers associated with status and control functions, and including the instruction address register (the program counter). A load/store architecture is used, and byte, signed and unsigned halfword, and word data types are supported.

The instruction set uses 16-bit and 32-bit formats (figure 41), in order to improve code density over the fixed 32-bit instruction length of the 801. To fit the instructions into 16 bits, a number of compromises are necessary. The size of the register bank is smaller than the 801, and many instructions use a two address form rather than a three address form. The register fields are in fixed positions, and the two operand registers are always fetched in parallel with instruction decode. One or both may be discarded, but they are fetched anyway in case they are needed.

The instruction set includes load and store multiple register instructions, which may transfer one to sixteen of the general registers. All load and store instructions use base plus immediate offset addressing; 16-bit forms have 4-bit offsets, 32-bit forms have 16-bit offsets.

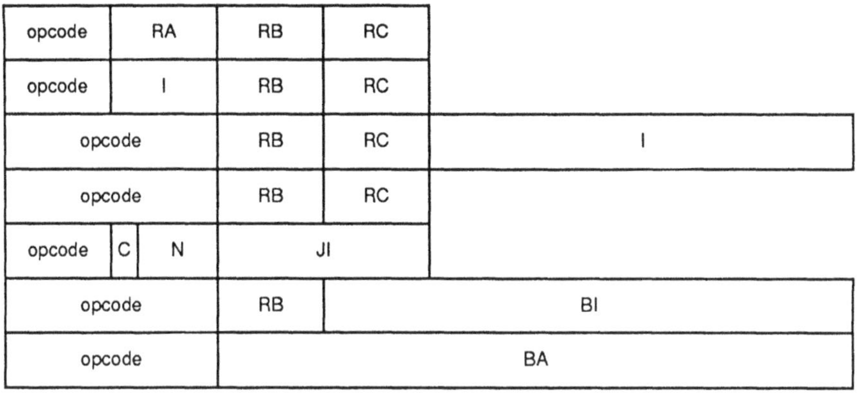

opcode	RA	RB	RC	
opcode	I	RB	RC	
opcode		RB	RC	I
opcode		RB	RC	

opcode	C	N	JI

opcode	RB	BI

opcode	BA

Figure 41: ROMP instruction formats
(From IBM, 1986)

16-bit jump instructions have a range of 254 bytes either way, and 32-bit branches have a range of 1 Mbyte either way. Branch and link instructions copy a return address into a register for subroutine linkage. An explicit branch with execute instruction implements a delayed branch.

3.1.2 ROMP Organization

The internal organization of the ROMP chip is shown in figure 42. The control is implemented in a small microcode ROM. The datapath operation takes three cycles:

(1)   Instruction decode/operand fetch.

(2)   Execute.

(3)   Write result back to register.

The operand fetch takes place during the second half of the cycle, whereas the result write takes place during the first half of the cycle. Therefore a register which is both written and read in the same cycle will give the correct value, and forwarding logic is only required directly around the ALU for the case where an instruction uses the result of the immediately preceding instruction.

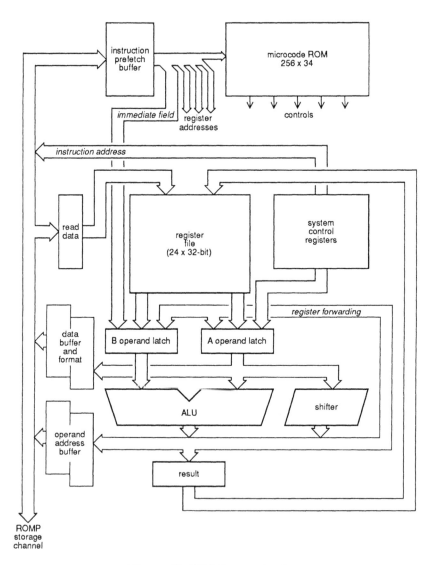

Figure 42: ROMP organization
(Adapted from IBM, 1986)

The ROMP storage channel handles all instruction and data memory references, and may have several outstanding requests at a time. Each request is accompanied by an identification tag, and returning values are matched by their tag to the appropriate destination.

The instruction prefetch buffer can hold up to four words of instructions, and will request a further word whenever it has space and there are no other requests for the storage channel.

The other end of the storage channel is supported by the Memory Management Unit (MMU) chip, which acts as an address translation buffer and a write buffer.

Floating-point support is provided by an adaptor which allows the connection of an NS32081 floating-point accelerator as a memory-mapped peripheral.

3.1.3 Conclusions

The ROMP architecture is unusual for a RISC in its use of two address operations and half-word instruction formats. The datapath organization and pipeline operation are similar to other RISCs, but the use of microcode instead of hardwired instruction decoding is not typical. The storage interface is quite sophisticated compared against the general simplicity of much of the design.

The design of ROMP diverged from the 801 because of the dependence of the 801 on high speed caches. The ROMP was developed over an atypical period of computer history (circa 1980), when VLSI RISC CPUs operated naturally with clock periods similar to main memory access times. VLSI CISC CPUs were slower than memory systems at this time. Under these circumstances, caches do not offer great performance improvements.

In most technologies and at most times CPUs are potentially much faster than main memories, and that is now true of VLSI RISC processors. Therefore the trade-offs applied to the ROMP architecture apply differently today, and the optimizations made on the 801 towards the efficient exploitation of caches are more likely to be relevant to any new design than those made on the ROMP towards direct connection of the CPU to main memory.

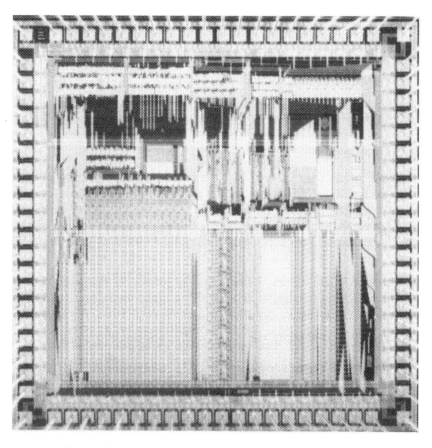

Figure 43: Photomicrograph of the Acorn RISC Machine
(Copyright ©Acorn Computers Limited, 1988.

3.2 THE VL86C010 ACORN RISC MACHINE

The majority of commercial RISC developments have tended to focus on maximizing the performance of the CPU at a cost level optimized for high-end engineering workstation or minicomputer applications. The RISC work at Acorn was initially targetted at PC applications, where the cost of the CPU must be much lower (though subsequent developments have allowed the same processor core to be applied at the higher cost/performance points as well; see the description of ARM3 in chapter 5).

The ARM uses 25,000 transistors on a 5.5mm square die. It is implemented on a 2 micron double-level metal single polysilicon CMOS process, and uses standard foundry design rules as supplied by VLSI Technology Inc., of San Jose, California. With a 12MHz clock it delivers around 6 MIPS sustained performance. VLSI Technology fabricate the device, and market both it and the associated chipset of support devices. The ARM is also suitable for use as a macrocell in a custom microcontroller.

The associated chipset consists of a memory controller, a video controller, and an I/O controller. With the addition of ROM, DRAM, and suitable peripheral controllers, these enable a high performance highly integrated low cost computer system to be constructed, such as the Acorn Archimedes range of personal workstations.

3.2.1 Architecture

The register organization on ARM is shown in figure 44. There are sixteen registers available at any time, of which fifteen are general-purpose and the sixteenth contains both the program counter and the program status register. Some of the registers are banked for supervisor and interrupt modes. There are two dedicated registers for supervisor mode, one for the return link and one for the supervisor stack pointer. Other registers may be saved by the supervisor code onto the supervisor stack, and then used as supervisor workspace.

The normal interrupt mode also has two dedicated registers, for similar reasons. The fast interrupt mode has seven dedicated registers, to allow I/O data transfers to be implemented in fast interrupt software without any register saving.

The program counter was included in the user accessible register set for many reasons. It enables the PC to be manipulated by standard instructions, for instance it can be reloaded on procedure return, and it can be used as a base for load and store instructions allowing position independent code to be

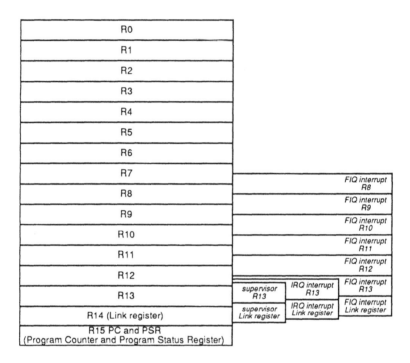

Figure 44: The ARM register organization

generated easily. Register 15 contains the PC and the status bits (figure 45), and these can be treated together when saving or restoring state, or separately when using the PC as a base address.

The instruction formats used by ARM are shown in figure 46. ARM has an unusually high number of instruction formats for a RISC processor, but the formats have considerable commonality in the positions of fields. The matching register fields are obvious from the figure; less obvious is the correspondence between (for instance) the OpCode field in the Data Processing format and the P and U bits in the various Data Transfer formats, but the commonality is still there. These fields control the function of the ALU during particular cycles, and their actions are performed by common hardware for all these instruction types.

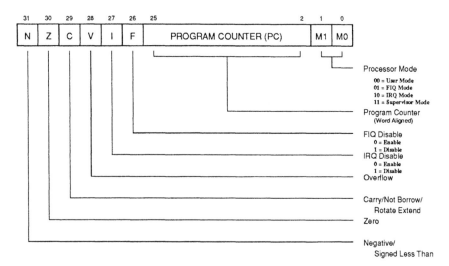

Figure 45: The ARM program counter and program status word

Though the register specification fields are in regular positions, some instruction decoding is required to locate the appropriate field for each bus in any cycle. This means that ARM cannot perform operand fetching at the same time as instruction decoding, which many RISC processors do. The relatively small register bank on ARM means that the operand fetch time is short, so the penalty associated with the serial access is small, but it is still a penalty. The benefit in instruction format flexibility is hard to weigh against this drawback.

The ARM instruction set is based on a load/store model. This means that no memory-to-memory operations are supported; all operands must be loaded into registers before they can be operated on. One class of instructions simply performs arithmetic and logical operations on register contents. A second class is concerned only with loading a value from memory or saving it back. A third class performs loads or saves of multiple registers, which has an obvious use in context changes, and less obvious uses in procedure entry and return and data copy operations. The other three instruction classes support branches, supervisor calls, and external coprocessors.

The entire instruction set is conditionally executable, so that short forward branches are usually unnecessary. Compilers can generate IF .. THEN ..

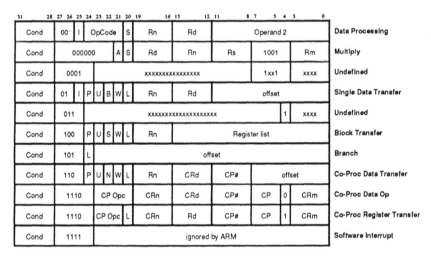

Figure 46: ARM instruction formats

ELSE code without any branches at all.

The instruction set is described in detail in the next chapter, where implementation details are also introduced. It is summarized in table 6.

All the ARM data operations (except the multiplies) allow an arbitrary shift of one of the operands before it enters the ALU, so there are no separate shift instructions.

3.2.2 Organization

The implementation of the processor is based around a 32-bit datapath (figure 47) which incorporates 27 registers, an ALU and a barrel shifter. An address incrementer generates the address during sequential instruction fetches and during multiple register transfer instructions. The ALU is used for address calculation when a branch or a data transfer requires a non-sequential location. One ALU operand always passes through the barrel shifter, so an operand alignment can be performed in addition to the ALU operation in every cycle. (For instance, the contents of a register can be multiplied by 5 in a single cycle by shifting it left two bits and adding it to itself.)

Table 6. The ARM instruction set

Data Operations

Producing a result:

MOV, ADD, SUB, AND, EOR, ORR
MVN   *(move not)*
ADC   *(add with carry)*
SBC   *(subtract with carry)*
RSB   *(reverse subtract)*
RSC   *(reverse subtract with carry)*
BIC   *(bit clear)*

Producing no result, used to set condition codes:

CMP   *(compare)*
CMN   *(compare negated)*
TST   *(bit test; equivalent to AND)*
TEQ   *(test equal; equivalent to EOR)*

Multiplies:

MUL   *(least significant 32 bits of a 32 x 32 product)*
MLA   *(multiply accumulate; as MUL, with the result initialized)*

Single Register Transfer

LDR   *(addressing modes are base plus index, or base plus offset;*
STR   *both with optional auto-index. Byte or word transfers.)*

(continued)

Table 6. (continued)

Multiple Register Transfer

LDM	*(any register subset; base addressing only, with optional*
STM	*auto-index. Stacks up or down memory, full or empty.)*

Branch and Subroutine Call

B	*(branch to anywhere in address space)*
BL	*(copy PC into R14 for return address)*

System Call Instruction

SWI	*(software interrupt)*

Coprocessor Instructions

CPDO	*(coprocessor internal data operation)*
CPDT	*(coprocessor multi-word data transfer to or from memory)*
CPRT	*(coprocessor single word register transfer to or from ARM)*

The ARM pipeline operates in three stages:

(1)   Instruction fetch.

(2)   Instruction decode.

(3)   Execute.

Each stage takes one cycle of the two phase non-overlapping clock, and a new instruction can start every cycle. The execute phase includes register read, shift, ALU and register write.

Because there is a single port to memory, a load or store instruction will take at least two cycles (one for the data transfer, and one for another instruction

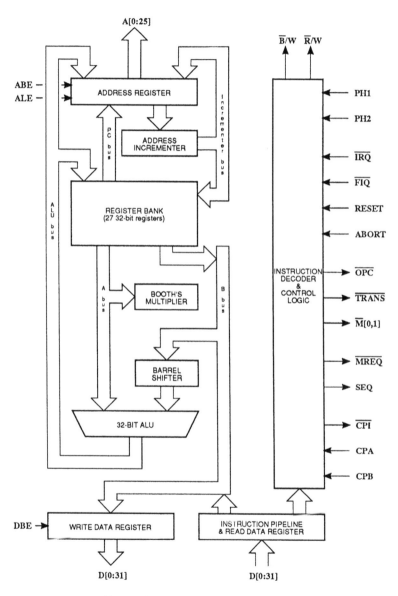

Figure 47: ARM block diagram

fetch). Therefore the store with base plus index addressing can be performed with a register bank with only two read ports.

3.2.3 The Memory Controller

The memory controller (MEMC) is a complete interface between ARM and up to 32 DRAM chips. It performs memory management functions (address translation and protection), and drives the multiplexed memory address lines and the timing strobes. It operates standard DRAM parts at their specified performance limits. No delay is introduced for address translation because this is performed during the DRAM precharge time for non-sequential cycles, and is unnecessary for sequential cycles. A single bank of DRAM, 32 bits wide, delivers 25 Mbytes/s of useable bandwidth when controlled by MEMC. The chip employs 30,000 transistors in a 5.5mm square of silicon.

The Memory Management Model

The memory management function of MEMC is performed by a content addressable memory (CAM) element. Main memory is divided into 128 physical pages, and each page has an entry in the CAM which contains the virtual address of the contents of that page. A virtual address is presented to the CAM at the start of each non-sequential memory access, and is compared with all the CAM entries in parallel. If a match is found, the CAM produces the required physical page number and the access completes. Otherwise the required data is known to be absent from memory, and the processor traps. In a demand-paged virtual memory system the software would normally locate the missing page on disk, transfer it into main memory, and then restart the aborted instruction. All the page table formats and paging algorithms are at the discretion of the system software.

This implementation of memory management is very much simpler than the translation cache mechanism which is frequently used by other manufacturers. It also obviates the need for a connection from MEMC to the main data bus, which saves pins on MEMC and reduces the capacitive load on the bus. (Registers on MEMC are programmed by encoding information onto the address bus.) It is very similar to the first virtual memory system, which was implemented on the Atlas computer at Manchester University in 1961 (and is described in the first chapter of this book).

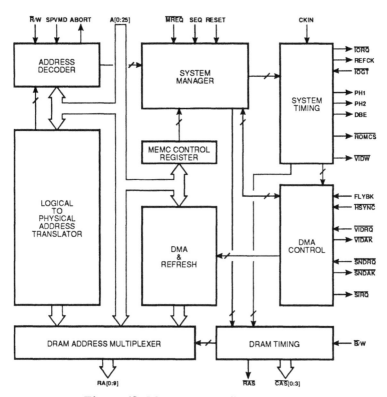

Figure 48: Memory controller block diagram

The Dynamic Memory Interface

The ARM is optimized for low cost personal computer applications, and in such systems the memory is usually a significant cost item. The only practical way of building memory for such systems is to use standard dynamic random access memory devices (DRAMs). They all have random access modes which allow about 4 million accesses per second, but they also have access modes which, though capable of at least twice the data rate, allow access to only a limited subset of locations. ARM has a control signal which indicates when the address for the next cycle will come from the address incrementer, and will therefore be sequential to the current address. This enables the memory system to select the fast mode when appropriate on a cycle by cycle basis, and with typical programs the memory system can use the fast mode for between 75% and 90% of all transfers. This increases the useable memory bandwidth by over 50% for no extra cost, which is important because memory bandwidth is the single most important determinant of processing performance.

Two other custom chips complete the ARM chip set. They are a video controller (VIDC) which generates a display using data from main memory, and an I/O controller (IOC) which allows standard peripheral devices attached to a separate I/O bus to be interfaced to ARM. The complete system organization is shown in figure 49.

The functionality of the processor may be extended by the addition of up to sixteen coprocessors, which follow the instruction stream on the main bus and cooperate in the execution of coprocessor instructions. One such coprocessor performs IEEE standard floating-point operations on single-, double- and extended-precision (80-bit) operands. If the coprocessor is absent the instructions trap, and may be emulated by software. The software emulator and the hardware coprocessor can be fully object-code compatible with each other.

3.2.4 Conclusions

The ARM processor has been optimized for close coupling to dynamic memories. This has influenced the design at many levels. Because of the relatively low speed of access of DRAMs, the entire execution phase (register read, shift, ALU, write) can operate in one cycle. This removes the need for complex forwarding logic to avoid data race hazards. The long cycle also allows more time for instruction decoding, which in turn allows more efficient use of the instruction bits, and more formats.

ARM is memory limited. Almost every memory cycle is used for an instruction fetch or a data transfer, so speeding up the execution unit by increasing the pipeline depth would have no benefit unless the memory could also be sped up. The most important issue is to make the best use of the fast local memory access modes which are used for sequential addresses; hence the condition field on all instructions (to avoid branches) and the multiple register transfer instructions (to minimize switching the address back and forth between instruction and data streams).

Because of the small die area, ARM is a very low cost RISC processor. It is therefore suited to PC, low-cost workstation and embedded controller applications. It is also uniquely suited to use as a processor cell on a larger ASIC design, and in addition to custom microcontrollers, this approach

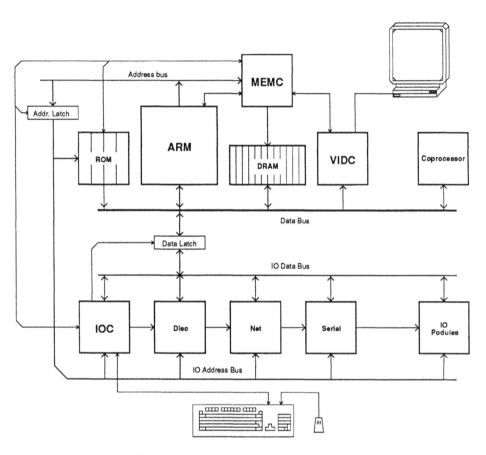

Figure 49: ARM system block diagram

allows a higher performance CPU to be built by the addition of on-chip cache memory (the ARM3, described in chapter 5).

In the next chapter of this book we shall look into the design of the ARM in some detail; the instruction set, implementation and design rationale will be further discussed there.

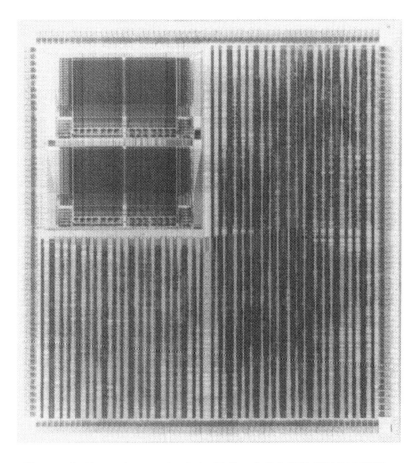

Figure 50: Photomicrograph of the Fujitsu MB86900 SPARC CPU
(Copyright ©Fujitsu.
Reprinted with permission of FUJITSU Mikroelektronik GmbH.)

3.3 THE SUN SPARC

The Sun SPARC (for *Scalable Processor ARChitecture*) is most novel for the way the business of processor development has been approached. Sun Microsystems has rapidly become established as a major manufacturer of workstations, which were initially based on the Motorola 68000 family of CISC microprocessors. They moved into RISC by defining an architecture, but they did not themselves implement that architecture. Instead, they licensed a number of independent suppliers to implement the architecture, and to market the implementation openly. The implementations are optimized at a variety of different cost/performance points.

The Sun approach emphasizes the distinction between *architecture* and *implementation*. The architecture defines the features of the processor which affect the logical result of a program; the implementation affects the performance and cost. A goal of the SPARC development is to define a single architecture which allows many implementations. The flexibility is underlined by the fact that the number of CPU registers is an implementation variable.

The MB86900 implementation of the SPARC architecture is based on a Fujitsu CMOS gate array. With a 16.67MHz clock it delivers around 10 MIPS sustained performance.

3.3.1 Instruction Set

The SPARC architecture defines three basic instruction formats, which are shown in figure 51. The 30-bit word offset of the CALL instruction allows a procedure to be called anywhere in the 32-bit virtual address space, whereas the branch instructions have a more limited range (though they can still reach any location within 8Mbytes). Most instructions take two source operands, one of which is always a register, and the second of which is either a register or a 13-bit sign extended immediate value. These two operands are combined in the specified way, and the result is written to a third register.

Nine data types are supported by the architecture. These are signed and unsigned bytes, halfwords, and words, and single, double, and extended floating-point values. The extended precision values use 80 bits of information, but require four words of memory.

A summary of the instruction set is presented in table 7.

The memory transfer instructions use base plus index or base plus 13-bit signed immediate offset addressing, with no autoindexing. The Fujitsu

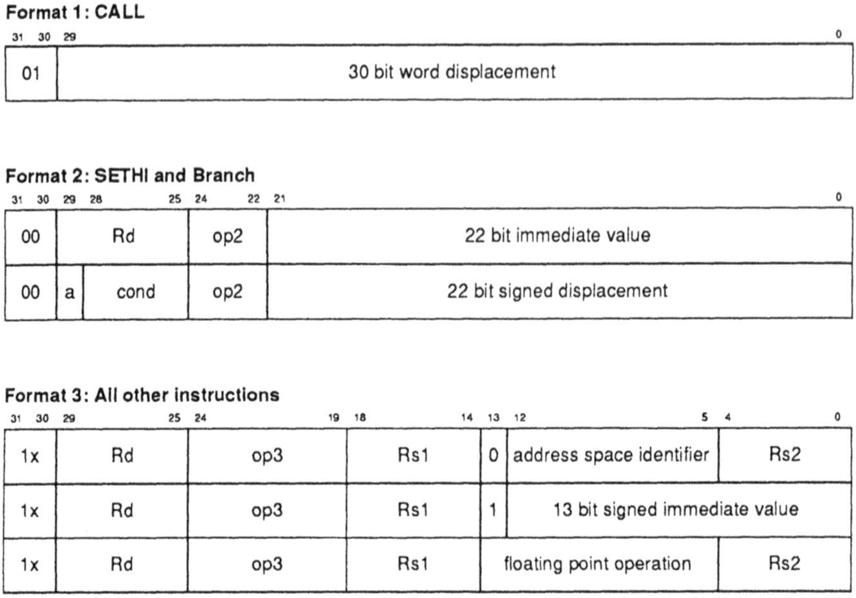

Format 1: CALL

31 30 29   0

01	30 bit word displacement

Format 2: SETHI and Branch

31 30 29 28   25 24   22 21   0

00		Rd	op2	22 bit immediate value
00	a	cond	op2	22 bit signed displacement

Format 3: All other instructions

31 30 29   25 24   19 18   14 13 12   5 4   0

1x	Rd	op3	Rs1	0	address space identifier	Rs2
1x	Rd	op3	Rs1	1	13 bit signed immediate value	
1x	Rd	op3	Rs1	floating point operation		Rs2

Figure 51: SPARC instruction formats
(From Sun, 1987)

MB86900 implementation achieves this with a dual ported register file by using multi-cycle load and store instructions. Other implementations may adopt different approaches.

3.3.2 Architecture

The SPARC architecture is based on a windowed register bank similar to the Berkeley RISC designs. The total number of registers may vary between one implementation and another. The Fujitsu MB86900 implementation has 120 general-purpose registers, organized as shown in figure 52. There are seven windows available here. The architecture allows up to 32 windows, and an implementation of the full 32 would require 520 registers in all.

The current window is changed by SAVE and RESTORE instructions, which are used on procedure entry and return. A trap also causes a new window to be entered (and RETT is the corresponding return). The flexibility of the architecture results from the transparent way that window overflows can be handled. The WIM (window invalid mask) register contains one bit for each

Table 7. The SPARC instruction set

Arithmetic and Logical Instructions

ADD/SUB	*(with or without carry, optionally setting condition codes)*
TADDCC	*(tagged add, set condition code, optional trap on overflow)*
TSUBCC	*(tagged subtract, set condition code, optional trap on overflow)*
MULSCC	*(multiply step and set condition codes)*
AND/OR	*(optionally setting condition codes)*
ANDN	*(logical inverse of AND)*
ORN	*(logical inverse of OR)*
XOR	*(optionally setting condition codes)*
XNOR	*(logical inverse of XOR)*
SLL	*(shift left logical)*
SR	*(shift right arithmetic or logical)*
SETHI	*(set the high 22 bits of a register)*

Data Transfer Instructions

LD	*(load single or double word, signed or unsigned byte, halfword)*
LDF	*(load single or double floating-point register)*
LDFSR	*(load floating-point status register)*
LDSTUB	*(load and store unsigned byte - for semaphores)*
ST	*(store single or double word, or byte or halfword)*
STF	*(store single or double floating-point register)*
STFSR	*(store floating-point status register)*
SWAP	*(swap a register with memory)*

Special Instructions

IFLUSH	*(instruction cache flush; traps when active)*
UNIMP	*(unimplemented instruction)*

(continued)

Table 7. (continued)

Control Transfer Instructions

BICC	*(branch on integer condition code)*
TICC	*(trap on integer condition code)*
FBFCC	*(branch on floating condition code)*
CALL	
RETT	*(return from trap)*
JMPL	*(jump and link)*
SAVE	*(decrement current window pointer to save register window)*
RESTORE	*(increment current window pointer to restore register window)*

Control Register Instructions

RDY/WRY	*(read/write the Y multiplication register)*
RDPSR/WRPSR	*(read/write the PSR)*
RDTBR/WRTBR	*(read/write the transfer base register)*
RDWIM/WRWIM	*(read/write the window invalid mask register)*

Floating-Point Operations

FABS	*(absolute value)*
FADD	*(floating-point add, single or double precision)*
FCMP	*(floating-point compare, single or double precision)*
FCMPE	*(as FCMP, with exception if unordered)*
FDIV	*(floating-point divide, single or double precision)*
FMOV	*(floating-point move, single precision only)*
FMUL	*(floating-point multiply, single or double precision)*
FNEG	*(floating-point negate, single precision only)*
FSUB	*(floating-point subtract, single or double precision)*
FaTOb	*(convert between integer, single and double precision formats)*

(Source: Sun, 1987)

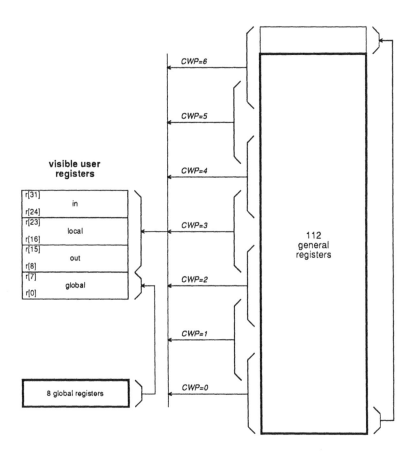

Figure 52: SPARC register organization

of the 32 windows defined by the architecture, and this bit defines whether or not the corresponding window is valid. An implementation with fewer than 32 windows will mark the unimplemented windows as invalid. The available windows may then be operated as a circular buffer, with a single window marked as invalid to break the circle. Procedure calls may go up and down the windows so long as the invalid window is not touched. If it is touched, a trap is caused, and one or more windows will be saved by the system before resuming with the invalid window pushed round the circular buffer a corresponding number of slots.

The *Current Window Pointer* (CWP) is contained in the program status register (figure 53), and has 5 bits for all implementations. The MB86900 does all operations modulo 7 to the CWP bits, to ensure that it never points to an unimplemented window.

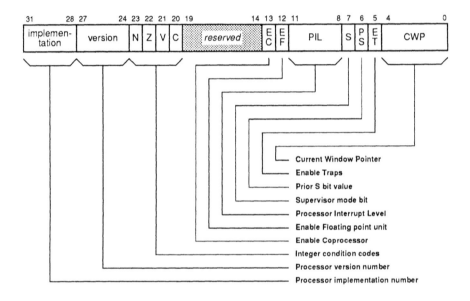

Figure 53: SPARC program status register
(From Sun, 1987)

The PSR also contains the integer condition codes which are optionally set by the data operation instructions and tested by the conditional branches. There are current and previous supervisor mode bits. The latter is used to record the mode before entry into a trap handler in order to allow it to be restored upon return.

Memory Management

A reference memory management architecture has been proposed for SPARC systems (Sun, 1988). As for SPARC itself, only the essential features are specified in the architecture definition. These include the page table entry format, the MMU control registers and the structure of the page table itself (which is conventional apart from the choice a three-level table rather than the more usual two levels). The organization of the TLB is left to the implementation; it may be of any size and any form.

3.3.3 Implementation

Though the instruction set is fully defined by the architectural specification, many details are a function of the implementation. For instance, the number of cycles taken by an instruction is not specified. Therefore the architecture allows simple implementations with a single ALU which use several cycles for an instruction, and it allows much more complex implementations with multiple function units which could perform several instructions in one cycle.

The first implementation of the SPARC architecture was the Fujitsu MB86900, which is based on a CMOS gate array. The internal organization is illustrated in figure 54.

This implementation has two function units; one for address calculations for load and store operations, and one for data processing. The data processing ALU includes a shifter which may be used to produce the result, or an arithmetic or logical result may be produced as normal. It is not possible to perform a shift in the same cycle as an arithmetic or logical operation, except by using the limited shifting capabilities of the aligner.

The SPARC instruction set includes floating-point instructions, but these are not implemented on the MB86900. A bit in the PSR may be cleared to cause every floating-point instruction to trap to emulation software, or an external hardware coprocessor may be attached. The MB86910 floating-point controller is such a coprocessor, and it uses the Weitek 1165 floating-point ALU chip and the 1164 floating-point multiplier chip to perform the operations. The floating-point operations proceed concurrently with integer activity, except for memory transfers where the two units must synchronize.

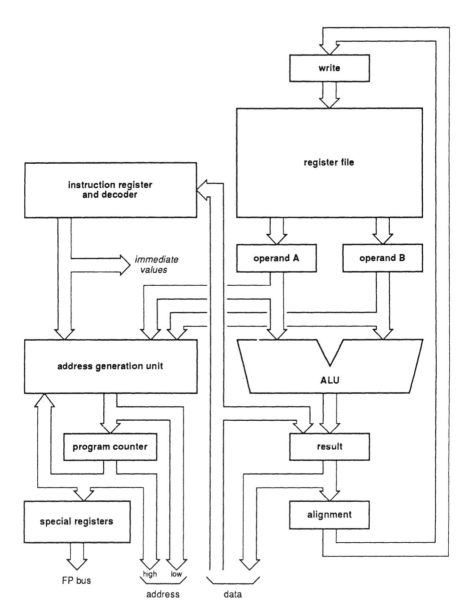

Figure 54: Fujitsu MB86900 internal architecture
(From Fujitsu, 1987)

3.3.4 Conclusions

The most significant features of the SPARC are:

- The clear distinction made between architecture and implementation, and the encouragement by Sun Microsystems of several third party vendors to produce implementations at different cost/performance points.

- The SPARC architecture is the purest commercial application of the original Berkeley register windows feature.

Only the MB86900 implementation of the SPARC has been described here. This is based on gate array technology, which is the fastest way to implement a new design. Higher performance variants are being built on CMOS standard cell, fully customized CMOS, and bipolar technologies.

3.4 THE MIPS R2000

All of the CPUs described in this chapter (with the possible exception of the Inmos transputer) contain ideas from the research processors described in the previous chapter, but their lineage is not always clear. The lineage of the MIPS R2000 is, however, perfectly clear. This processor is a direct descendant from the work at Stanford; the company name is the name originally given to the Stanford CPU, the architectural principles are those developed during the Stanford research, and the design team included many principals from the Stanford project.

The R2000 architecture may therefore be viewed as a second generation of the Stanford approach. The differences between the generations should provide an insight into the lessons learnt from the the first version. For a full description of the R2000 family the reader is referred to Kane (1987).

The R2000 is built on a 2 micron double-metal CMOS process, and uses 110,000 transistors. With a 16.67MHz clock it delivers around 12 MIPS sustained throughput.

3.4.1 The R2000 Architecture

The first notable difference between the Stanford MIPS architecture and that of the MIPS R2000 is in the instruction set. The Stanford processor allowed two instructions to be encoded in one 32-bit word, whereas all R2000 instructions are exactly one 32-bit word. This is probably a direct consequence of initiating a new instruction every cycle, rather than every other cycle as in the older design.

The number of general-purpose registers has also been increased, to 32. The programmers view of the CPU registers is shown in figure 55. The two special registers dedicated to multiplies and divides remain, and the program counter is a separate register also. Note that like the Stanford MIPS, the R2000 has no condition codes.

In addition to the integer CPU, the R2000 chip also contains the *System Control Coprocessor*, which performs address translation for virtual memory support and assists the CPU to recover from exceptions. This unit has additional programmer visible state, including the processor status register, and will be described later.

The instruction set uses three basic instruction formats (figure 56), all of which occupy exactly one word. Basic arithmetic and logical operations use three register addresses, and memory references are restricted to LOAD and

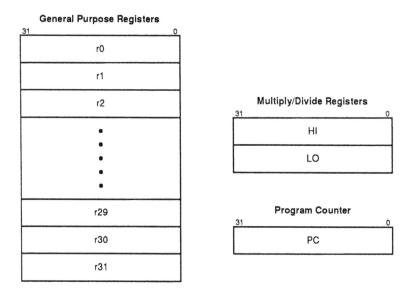

Figure 55: R2000 CPU registers
(From Kane, 1987)

STORE instructions with base + 16-bit immediate offset addressing.

The full instruction set is summarized in table 8. The following are additional notes adding some details:

- The load/store word left and load/store word right instructions are used in pairs to access misaligned words in memory.

- Any 32-bit number can be loaded into a register in two instructions: a LUI to load the high halfword and clear the low halfword, followed by an ADD unsigned (or an OR) immediate to load the low halfword.

- Any memory address can be accessed by using a LUI to load a base register and then using the 16-bit immediate offset in the LOAD instruction itself to complete the address.

- The CPU can be configured to operate with big-endian or little-endian addressing at reset. This only affects the operation of LOAD and STORE instructions.

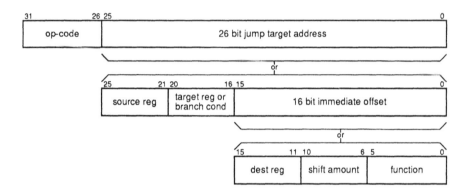

Figure 56: R2000 instruction formats
(From Kane, 1987)

- The conditional branches perform the comparison as part of the instruction; no separate condition codes are kept. 'Set on less than' instructions place a logical result in a general register.

- Signed ADD and SUB trap on arithmetic overflow. Unsigned ones do not, nor do multiply and divide under any circumstances.

Hardware Interlocks

The R2000 inherits from the Stanford MIPS an architecture where the compiler has to organize code to avoid data race hazards. This applies particularly to LOAD instructions, where the instruction following the load will get the wrong value if it uses the target register. Many processors support delayed loads, but most have hardware interlocks which will stall the processor if an attempt is made to use the target register before it is valid. The R2000 has no such interlock, and the compiler must either find a useful instruction which does not use the loaded value to follow the load (which is usually possible), or else insert a no-op.

The integer multiply and divide instructions do have hardware interlocks. These are executed in an autonomous unit, and other instructions can continue in parallel, but an attempt to read the result from HI or LO will cause the processor to stall until the result is ready. The interlocks are important here, as these instructions take several cycles to complete, and it would often be hard to find sufficient useful work to do in parallel.

Table 8. The R2000 instruction set

Arithmetic and Logical Operations

ADD	*(signed or unsigned, register or 16-bit immediate 2nd operand)*
SUB	*(signed or unsigned, register second operand only)*
SLT	*(set on less than, signed/unsigned, register or 16-bit immediate)*
AND/OR	*(register or 16-bit immediate second operand)*
XOR	*(register or 16-bit immediate second operand)*
NOR	*(register second operand only)*
LUI	*(load upper 16 bits with 16-bit immediate, zero lower 16 bits)*
S	*(shift left/right, arithmetic/logical, by immediate/register)*

Multiply and Divide Instructions

MULT/DIV	*(signed or unsigned)*
MTHI,MFHI	*(move to and from HI register)*
MTLO,MFLO	*(move to and from LO register)*

Loads and Store Instructions
(all use base plus 16-bit immediate offset addressing)

L	*(load word, or signed or unsigned byte or halfword)*
LWL,LWR	*(load left or right section of misaligned word)*
S	*(store word, byte or halfword)*
SWL,SWR	*(store left or right section of misaligned word)*

Special Instructions

SYSCALL	*(system call)*
BREAK	*(breakpoint trap)*

(continued)

Table 8. (continued)

Program Flow Control Instructions

J	*(jump by replacing bottom 28 bits of PC with immediate value)*
JAL	*(jump and link: as jump, but saving a return address in r31)*
JR	*(jump to address in register)*
JALR	*(jump to address in register;*
	save return address in specified register)

branches (16-bit PC relative signed word offsets):

BEQ/NE	*(if specified registers are Equal/Not Equal)*
B..Z	*(if specified register is Greater/Less than (or Equal to) Zero)*
B..ZAL	*(if register Greater/Less than Zero; save return address in r31)*

Coprocessor Instructions

LWCz	*(load word from coprocessor z)*
SWCz	*(store word to coprocessor z)*
MTCz/MFCz	*(move to/from coprocessor z)*
CTCz/CFCz	*(move control to/from coprocessor z)*
COPz	*(coprocessor z operation)*
BCz	*(branch 16-bit PC relative on coprocessor z True/False)*

System Control Coprocessor (CP0) Instructions

MTC0/MFC0	*(move to/from coprocessor 0)*
TLBR/TLBWI	*(read/write indexed TLB entry)*
TLBWR	*(write random TLB entry)*
TLBP	*(probe TLB for matching entry)*
RFE	*(restore from exception)*

(Source: Kane, 1987)

3.4.2 The System Control Coprocessor

The System Control Coprocessor is part of the R2000 chip, and performs memory management and exception handling functions. These two functions may be considered separately.

Memory Management is performed by a software-maintained fully associative on-chip translation lookaside buffer (TLB) and supporting registers. The TLB contains the virtual to physical address mappings for 64 4Kbyte pages, where the virtual address is the CPU's 32-bit address extended by adding a 6-bit process ID. When a virtual address is presented to the TLB and fails to match any of the 64 entries, an exception is raised. The exception handler must identify the required translation and copy it into one of the TLB entries before resuming the original task.

The address exception software may use any algorithm for determining which TLB entry to use next, but an on-chip pseudo-random number generator is the normal selection method. This is in fact a counter clocked by the CPU clock, but since this frequency is very high compared with typical page fault frequencies, the effect is random. The random number register will never produce an answer below 8, so TLB entries 0 to 7 may be allocated to system critical pages without fear of accidental removal (and without the overhead of checking on every exception that those pages are safe).

Software TLB management is flexible and simple to implement, but could be very inefficient if many instructions are required for each page fault. The System Control Coprocessor contains additional registers to help with exceptions, and particularly to make TLB updates efficient. The *Bad Virtual Address* register contains the most recent virtual address which caused a page fault. This can be used to find the required page table entry in main memory, but only after some manipulation. The *Context* register contains the top bits of the bad virtual address (which correspond to the bad virtual page number) appended to a writable page table base address. This will normally be exactly the pointer required to load the PTE without any manipulation.

Another System Control Coprocessor register contains the correct address to return to from the exception. This may seem obvious, but remember that the R2000 supports delayed branches, so the return address is not the obvious one if the faulting instruction occupies a branch delay slot. However the R2000 architecture has been designed so that the branch can always be executed again, so the exception return address will point to the branch if the instruction in the delay slot faults.

3.4.3 Organization

The organization of the R2000 chip is shown in figure 57 in block diagram form. Note that the CPU has a separate address adder in addition to the ALU; this is important to the operation of the pipeline. The multiply/divide unit is autonomous, and can operate concurrently with other datapath activity so long as the result is not required.

Pipeline Operation

An instruction is executed in five cycles:

(1)   IF - instruction fetch. The program counter address is translated and used to access the instruction cache.

(2)   RD - operand read, instruction decode. The source registers are accessed in parallel with instruction decoding.

(3)   ALU - ALU operation. The ALU (or the shifter) generates the result, or the address adder calculates the memory address.

(4)   MEM - access memory. The data cache is accesses if the instruction is a load or a store.

(5)   WB - write back result. The ALU/shifter result or loaded data value is written back into a register.

The basic operation of the R2000 pipeline is shown in figure 58. The internal operations correspond only loosely to the cycles named above, though the design of the physical pipeline is anything but loose.

To get some idea of how tight the pipeline design is, consider the following cases with reference to figure 59, where three instructions are shown in their correct overlapped positions:

(1)   If instruction 1 is a LOAD, it must fetch the data in time for its use as an operand by instruction 3 (since the architecture specifies a single delay slot after a LOAD). In cycle 3, instruction 3 uses the TLB during the first half of the cycle to translate the PC. Instruction 1 must use the TLB during the second half of the cycle in order to be ready to access the data cache in the following cycle. If the main ALU were responsible for address calculation, its result would be valid too late. The R2000 has a fast address adder (which only has to do register plus 16-bit addition) to get a valid address half a cycle earlier.

(2)   If instruction 1 is an unconditional BRANCH, it must calculate the target before the start of cycle 3 in order that instruction 3 can be

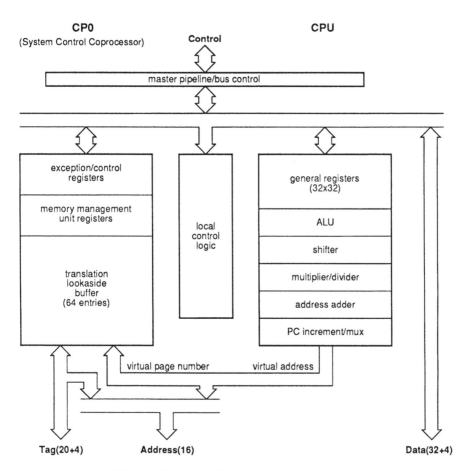

Figure 57: R2000 functional block diagram
(From Kane, 1987)

fetched from the correct address (BRANCHes only have a single delay slot in the architecture definition). This is only possible because the BRANCH instructions are all PC relative, and can therefore begin calculating the target address as soon as the instruction arrives. They do not need to access a general register, and the 16-bit fast adder can again be used to produce an early result.

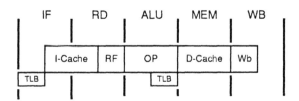

Figure 58: R2000 pipeline operation
(From Kane, 1987)

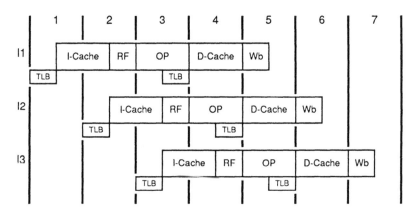

Figure 59: R2000 instruction overlap
(From Kane, 1987)

(3) A JUMP instruction may require a general register to be fetched, but it does not have to add in an offset, and again the address is ready just in time to be used for passing to the TLB at the start of cycle 3.

(4) A conditional BRANCH instruction can calculate the target in the same way as an unconditional BRANCH, and in parallel access the register(s) required to evaluate the condition. To get all the information ready in time to decide whether or not to send the target to the TLB at

the start of cycle 3 must require a very fast comparator in the case of BEQ and BNE!

(5)   Note how instruction accesses are half a cycle skewed with respect to data accesses. This allows them to share the fast address adder, the TLB, and the data and address bus (provided that all of these can operate twice in a cycle).

3.4.4 The R2010 Floating-Point Coprocessor

The R2000 architecture supports up to 4 coprocessors. The system coprocessor is on the R2000 CPU chip itself. The R2010 Floating-Point Accelerator (FPA) is a coprocessor on a separate chip. The basic structure of the R2010 is shown in figure 60. The FPA tracks the R2000 instruction stream and decodes instructions in step with the main CPU.

The floating-point logic is divided into three main functional units, handling floating-point add, divide and multiply respectively.

The three functional units can operate autonomously, only possibly clashing for resources at the beginning and end of operations. The common register set is a source of contention, if an operation requires the result of another operation which has not completed. Such data dependencies are handled by hardware interlocks which stall the unit until the result is ready. Another possible contention arises when the result is to be normalized because the add unit performs this part of all operations. In general it is possible for a compiler to operate the unit with considerable overlap of activity across the three functional units.

The R2010 is very closely coupled to the R2000 CPU. Its pipeline operation (figure 61) is the same except that the R2010 defers result writeback for one cycle so that the R2000 and R2010 can coordinate any exception response which might have arisen.

3.4.5 R2000 Cache Organization

As was noted above, the R2000 can produce one instruction address and one data address every cycle. These addresses, and the data responses, share the same pins on the CPU. In a high performance system (figure 62) they will be demultiplexed externally to operate separate data and instruction caches.

The CPU chip has logic to perform all the necessary cache control functions. The only external components which are needed are for the address demultiplexing logic, and RAM for the instruction and data stores and

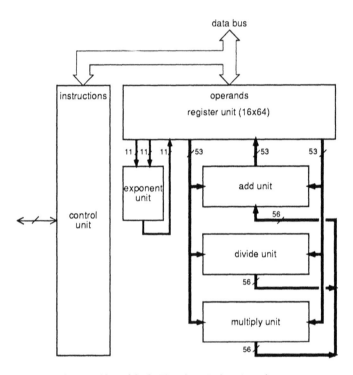

Figure 60: R2010 Floating-Point Accelerator
(From Kane, 1987)

associated address tags. The caches are both direct mapped, and physically addressed (via the on-chip TLB). The R2000 produces strobes for controlling the latches and enabling the RAM chips for reading or writing.

The low address lines (which are used for indexing into the cache) are sent out first, and the corresponding tag, data, and valid bits read onto the CPU. If the tag matches the on-chip high physical address, and the entry is valid, the access completes and the data is used. Otherwise an access must be made to main memory, and the processor will then drive the tag lines to update the cache entry when the data arrives.

The cache uses a write through strategy, which means that any write can simply update its corresponding cache entry (there is no question of first having to identify the correct entry in a direct mapped cache; a particular

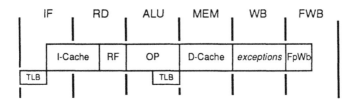

Figure 61: R2010 FPA pipeline operation
(From Kane, 1987)

address can only ever be stored in a unique location). To avoid the processor being held up while the data is copied to main memory, the R2020 write buffer chip copies writes to the cache and completes the transaction with main memory while the CPU continues with other instructions. The R2020 can queue up to four memory writes if the CPU issues them faster than memory can accept them.

3.4.6 Conclusions

The MIPS Computer Systems architecture started from the efficient but slightly quirky architecture of the Stanford MIPS machine. They took the best ideas and changed the others to produce an extremely clean design. Everything about the R2000 is balanced and tight.

The single delay slot pipeline design is probably optimal. Two slots can be hard to fill usefully, and zero delay is not possible with single cycle LOAD instructions. BRANCH delay slots can be avoided by intelligent instruction prefetch units, but they are far more complex and no more effective than the R2000 solution.

The absence of a hardware interlock on the LOADed value is a debatable feature. With or without it a compiler should try to find an independent instruction to put in the delay slot, but it would be reassuring to know that the interlock was there just in case of an error, especially when programming in assembler.

The absence of condition codes makes the code for some operations unusually long. 64-bit adds take four instructions, for example, whereas on many RISC processors two are sufficient. Such operations are infrequent, however, so the omission of efficient support for them is justified by the RISC criterion.

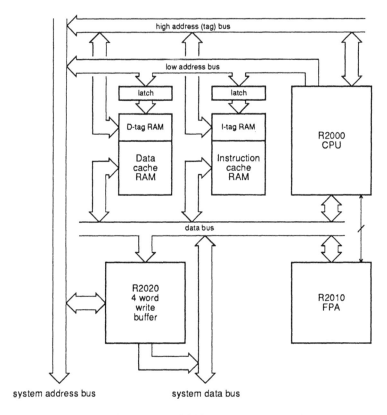

Figure 62: R2000 system organization
(Adapted from Kane, 1987)

Overall the R2000 is the epitomy of elegant RISC architecture. If maximum performance is the principal goal, the R2000 achieves it with great simplicity through careful matching of all the elements of the processor, caches, buses, pipelines, coprocessors, etc.

The R3000 series RISC architecture is a later development which is very similar to the R2000, but combines a higher maximum clock rate with a number of improvements such as more sophisticated control of the caches.

3.5 THE AT&T CRISP

The CRISP microprocessor was designed at AT&T's Bell Laboratories as part of an on-going research project to develop architectures which would execute the C language efficiently with a reasonably simple compiler. It is a 32-bit processor on a single CMOS chip, and uses 172,163 transistors on a 1.75 micron double polysilicon single metal layer process. It embodies a number of novel architectural features, as well as drawing on past experience from both RISC and CISC designs. With a 16MHz clock it is rated at about 10 MIPS.

Most notable amongst the novel features are the decoded instruction cache with branch folding, and the stack register cache. The design takes memory to memory operations, variable length instructions and a simple compiler model from the CISC approach, and adds pipelining, one instruction per cycle and a small number (four) of addressing modes from the RISC approach. This blend of old and new characteristics makes the CRISP a very interesting device.

3.5.1 Instruction Set

The processor understands only byte, halfword and word data types. All may be signed or unsigned, and the shorter types are sign extended where appropriate.

Addresses of data items may be immediate, absolute, stack pointer relative or stack pointer relative indirect. Jump and procedure calls may use absolute, absolute indirect, stack pointer relative indirect or PC relative addresses.

All data operations are memory-to-memory (where hopefully many of the memory references will hit the on-chip stack cache). Dyadic operations may have two fully specified addresses for the operands, and place the result in an implicit address (the *accumulator*, which is a fixed offset from the stack pointer). This is called the 2.5 address form. Or they may return the result to one of the operand registers (the 2 address form). A summary of the instruction set is given in table 9.

The instructions are encoded in main memory in 1, 3 or 5 halfwords. The first halfword contains full details of the required operation, and subsequent halfwords contain extended address information when required. The instruction length is encoded in the first halfword in a form which is very easy to decode, for easy Next-PC generation.

Table 9. The CRISP instruction set

Arithmetic and Logical Instructions

2 and 2.5 address forms:

ADD, SUB, MUL
QUO   *(division)*
AND, OR, XOR
SHR   *(arithmetic shift right)*
USHL   *(unsigned shift left)*
USHR   *(unsigned shift right)*

2 address forms:

UMUL   *(unsigned multiply)*
UQUO   *(unsigned divide)*
UREM   *(unsigned remainder)*
CMP   *(equality and signed and unsigned less than comparison)*
MOVE, MOVA   *(move effective address)*
ADDI, ANDI, ORI *(interlocked bitwise add, and & or)*

Stack Management Instructions

ENTER   *(allocate stack frame)*
RETURN   *(and deallocate stack frame)*
CATCH   *(restore stack from memory)*

Program Flow Control Instructions

JMP   *(unconditional, if true, false, carry, or overflow)*
CALL   *(procedure call)*

(continued)

Table 9. (continued)

System Call Instructions

KCALL	*(kernel call)*
KRET	*(kernel return)*
CPU	*(internal register access)*

(Source: Ditzel, McLellan and Berenbaum, 1987)

An interesting feature of the CRISP instruction set is the inclusion of a branch prediction bit, whereby the compiler can indicate statically whether a conditional branch is likely or unlikely to be taken. The compiler also tries to move the instruction which generates the condition on which the branch decision is based as early as possible before the branch; if it is far enough ahead the outcome of the branch can be determined by the hardware, but if this is not possible the prediction bit is used to follow the most likely path. Only when the prediction bit turns out to be wrong does the pipeline stall while instructions are fetched from the correct path.

It is interesting to compare this static branch prediction with the branch target cache of the AMD Am29000, which may be thought of as a dynamic branch prediction mechanism.

3.5.2 Architecture

The overall structure of the chip is shown in figure 63. There are three separate cache structures: The prefetch buffer holds instructions in the same variable length form as they are held in main memory; the decoded instruction cache holds instructions in a 192-bit wide decoded form as required by the execution unit, and the stack cache holds the data items which are at the top of the stack at present. All the caches are directly mapped by address.

In addition to the three caches the processor contains an I/O unit which handles communication with main memory, and two functional units. The functional units are autonomous without a central control unit. The prefetch/decode unit fetches instructions into the prefetch buffer and decodes

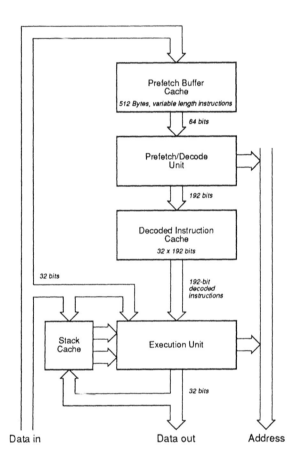

Figure 63: CRISP block diagram
(From Ditzel, McLellan and Berenbaum, 1987)

them into the decoded instruction cache. The execution unit processes the instructions and modifies user data structures accordingly.

There are two unusual consequences of this independent prefetching and decoding activity. Firstly, the prefetcher can see branches coming, and use the *hint* bit to determine whether the branch is likely to be taken. It can therefore follow the most likely execution path before the branch has reached the execution unit, and maintain the supply of instructions. The pipeline break problem is resolved without recourse to delayed branches, which is in line with the original philosophy of keeping the compiler simple.

Secondly, because the prefetch unit follows the branch when it is likely to be taken, the branching action required of the execution unit is minimal. The decoded instruction format allows a branch action to be specified simultaneously with an ALU operation, effectively allowing the branch to be *folded* into a concurrent instruction. A branch therefore adds no cycles to the execution time of a routine in the most common cases.

The Execution Unit

The execution unit consists of three pipeline stages (figure 64). The *Instruction Register* stage contains information about the operands, either the operand value itself (in the case of an immediate operand), or the address of the operand (absolute addressing and stack pointer relative) or the address which contains the operand address (indirect addressing). The *Operand Register* stage contains the actual operand values. The *Result Register* stage contains the result of combining the operands in the specified way. Each stage may be controlled by a different instruction at any point in time.

A particular instruction begins execution by moving from the decoded instruction cache into the Instruction Register. The operand address registers are added to the stack pointer for stack pointer relative direct or indirect addressing, and two 28-bit adders are dedicated to this function. Here any references to off-chip memory must be detected and the data items read in; indirect addressing requires two such memory accesses. Memory accesses which happen to use addresses currently contained in the on-chip stack cache must be detected and the values located. The values are moved into the Operand Registers, and when they have both been loaded the instruction moves on to the Operand Register stage and frees up the Instruction Register stage for the next instruction. Note that the stack cache is dual ported, so that if both operands refer directly to memory which is in the stack cache, then both operands can be located and fetched in one cycle. The instruction may only occupy the Instruction Register stage for one cycle even though it apparently requires two memory accesses to complete.

The instruction moves to the Result Register stage by processing the operands in the ALU. The values in the Operand and Result Registers are all 32-bit words, so byte and halfword values must be aligned (by shifting) and sign extended on their way into the 32-bit ALU, and a byte or halfword

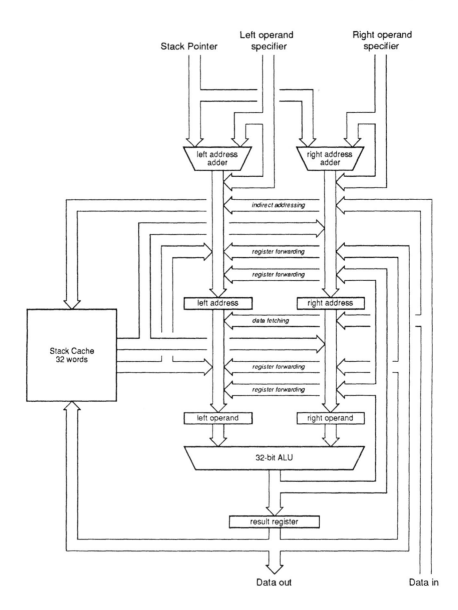

Figure 64: The CRISP execution unit
(Adapted from Ditzel, McLellan and Berenbaum, 1987)

result must also be aligned on the way out.

The pipelining of the execution unit may mean that a required memory value is not up to date, as a previous instruction may be constructing a new value for the same address further down the pipeline. CRISP has full hardware checks for this situation, and will forward the new value directly when it is available. This automatic register forwarding could have been omitted and left for the compiler to sort out, but it was included to satisfy the objective of keeping the compiler simple.

The Stack Cache

The stack cache is based on the Bell Labs C Machine architecture (Ditzel and McLellan, 1982). It is a 32-entry direct mapped cache of memory values which lie at addresses delimited by two dedicated registers: The *Stack Pointer* is the base register, and the *Maximum Stack Pointer* is one word on from the highest cached word. An address less than the Maximum Stack Pointer and greater or equal to the Stack Pointer is in the cache. The Stack Pointer is replaced by the *Interrupt Stack Pointer* for interrupt processing, which supports a separate stack for system use.

The stack cache is dual ported, and accesses values for both operands at the same time. It is implemented as two banks of single access RAM, one for each operand. When a new value is written into the cache it is copied into both banks. The use of two banks of standard RAM rather than one bank of specially designed dual ported RAM avoids the time and cost of designing the latter, and probably takes up very little extra space.

Whenever a memory data access is required, the address is compared with the Maximum Stack Pointer and the Stack Pointer. Both operand addresses may be compared simultaneously with both stack pointers, and either or both may be found to be within the stack. Both RAM banks are accessed, and both memory accesses complete in the one cycle when both are within the cache, which will be the normal case. A pipelined result can also be written back during the same cycle. In this way the stack cache allows register to register performance to be achieved with memory-to-memory instructions.

Cache management is performed explicitly by certain instructions:

- *Enter* is used on procedure entry to allocate a new stack frame, which requires the Stack Pointer to be moved. Since this may re-use some of the cache, those entries are flushed (rewritten back to main memory) appropriately.

- *Return* deallocates the stack frame (by moving the Stack Pointer back to where it was before the *Enter* instruction) and branches back to the address saved by the previous *Call* instruction.

- *Catch* restores from memory the cache entries flushed by the previous *Enter* .

The compiler has complete control over the section of memory which is kept in the cache, and can use this to minimize data traffic to and from external memory.

The Instruction Decoder

The instruction decoder is an independent processor in its own right, and is responsible for converting the variable length instructions in the prefetch buffer into fixed length instructions for the execution unit. Most of the processing of the program counter is performed here, which allows instruction prefetching and decoding to proceed ahead of the execution unit. The only PC related function of the execution unit is to choose between two alternative PC values (generated by the instruction decoder) when a conditional branch is executed. A bit in each conditional branch instruction is set by the compiler to indicate to the instruction decoder whether it should continue prefetching from the branch destination (branch taken) or from the current address (branch not taken). Only when the execution unit chooses the opposite route to that indicated in the instruction will the prefetch pipeline be broken, and the execution unit forced to wait while the correct instruction is fetched. Some mainframe computers prefetch from both possible paths when a conditional branch is found, but with the limited memory bandwidth available in a typical microprocessor application this is not usually practical, and the CRISP approach is a good compromise.

The decoded instruction format held in the Decoded Instruction Cache includes a *Next-PC* field and an *Alternate Next-PC* field. In normal execution the Next-PC field contains the address of the next instruction, which is calculated by the instruction decoder, and depends on the current PC and the length of the current instruction in undecoded form. If a branch follows the current instruction, it can be merged by the instruction decoder by modifying these two fields. The execution unit will perform the current instruction and the subsequent branch in the same time as the current instruction would have taken on its own. This branch folding results in branches taking no time up in the execution unit, and since branches typically represent 35% of all instructions this is no small saving.

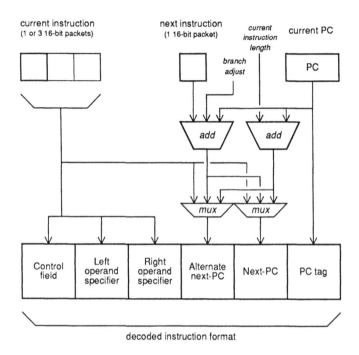

Figure 65: Branch folding on CRISP
(Adapted from Ditzel and McLellan, 1987)

The decoded instructions are placed into the decoded instruction cache, where they are stored until required by the execution unit. This cache decouples the instruction decoder from the execution unit, enabling both to operate at their own rate.

3.5.3 Conclusions

CRISP has several novel features with desirable properties, and as usual the value of such features depends on a cost/speed/complexity trade-off.

- The compact instruction format in main memory results in dense code, which reduces both the cost of main memory for a given program and the bandwidth requirements for a given execution speed.

- The instruction decoder expands the compact format into a simple form, enabling a RISC execution unit to deliver RISC performance. The drawback is the complexity added in the form of the decoder, and the increase in the pipeline length which increases branch latency when the branch prediction mechanism fails.

- The branch folding is a very good way of removing 35% of instructions from the execution unit, and other approaches to this important problem will doubtless appear in the future.

- The stack cache is a novel organization of the register set which avoids the problem of compiling onto a fixed register bank. The advantage is weakened somewhat now that techniques have been found for building compilers which make good use of a conventional register bank. The stack cache is also easily enlarged when a process with a smaller feature size is available, whereas a fixed register bank is hard to extend without a complete redesign of the instruction set. (The number of register windows may be increased in a Berkeley RISC style of architecture, but it is not usually possible to increase the number of registers visible at any one time.)

- The stack pointer relative addressing, in combination with the stack cache, gives high speed access to the on chip registers with the same number of instruction bits as would be required for conventional register access. This avoids the usual problem of memory-to-memory architectures, which sometimes require many instruction bits to specify the memory address.

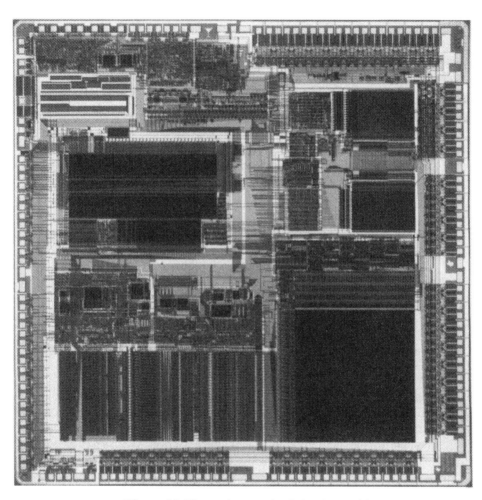

Figure 66: Photomicrograph of the Am29000
(Copyright ©Advanced Micro Devices, Inc., 1988.
Reprinted with permission of copyright owner. All rights reserved.)

3.6 THE AM29000

The Am29000 from Advanced Micro Devices is a single-chip 32-bit microprocessor with on-chip memory management and instruction cache. The chip design assumes the existence of an external high bandwidth instruction memory, and the instruction cache is activated only on branches, where it maintains the supply of instructions during the memory latency period.

The chip is characterized by a large general-purpose register file which is uniformly visible to the user. The registers may be treated as a stack cache, if suitable procedure entry and exit mechanisms are used. Most instructions execute in a single cycle, including loads, stores and branches. The pipeline uses hardware interlocks to manage data dependencies. The on-chip translation buffer is managed by software, so the translation table format is user-definable.

The Am29000 is implemented on a 1.2 micron CMOS process. It uses 210,000 transistors, and with a 25MHz clock it delivers a sustainable performance of around 17 MIPS.

3.6.1 Instruction Set

All Am29000 instructions have the same basic format. They are 32 bits long, and contain four byte-length fields (figure 67).

A summary of the instruction set is presented in table 10.

Features to note in the instruction set include:

- The field extraction and insertion instructions, which improve the efficiency of byte and half-word handling to compensate for the absence of explicit loads or stores of quantities less than a word (though it is possible to write a single memory byte with correct use of the memory interface signals).

- The use of general registers to hold the results of logical comparisons, instead of flags in a program status register.

- The 'assert' instructions, which cause traps when the assertion is false. This is a very efficient way of checking for errors which should not normally happen, such as stack overflow and array bound checks. A conventional instruction set would have to compare and branch conditionally to the error handler, taking two instructions. The assert and trap instruction may be relatively slow when the trap is taken, but

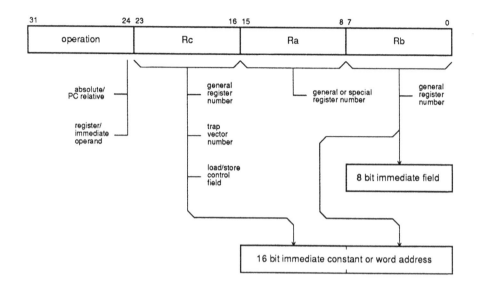

Figure 67: Am29000 instruction format
(Adapted from AMD, 1987)

this is normally irrelevant as it should not happen. What matters is that it is quicker under the usual condition when the trap is not taken.

- The arithmetic trap on overflow instructions allow safe code to be written without the overhead of explicit tests for overflow. This leads to safer code, since if the explicit tests have an overhead they are often omitted.

- The definition of complex instructions which currently trap allows for fully code-compatible implementations in the future when more of these functions may be implemented in hardware, either on-chip or in a coprocessor.

- The load and store instructions use very simple addressing modes, principally register indirect. All address calculations are done explicitly by data operations. This requires extra code, but enables single cycle load and store with a triple ported register file, and the calculated address can be re-used subsequently without the overhead recurring.

Table 10. The Am29000 instruction set

Arithmetic Operations

ADD   *(32-bit add, with/without trap on signed/unsigned overflow)*
ADDC   *(add with carry, with/without trap on signed/unsigned overflow)*
SUB   *(32-bit subtract, with/without trap on signed/unsigned overflow)*
SUBR   *(as SUB, with operands reversed)*

MUL   *(multiply step, signed and unsigned)*
MULL   *(multiply last step)*
MULTIPLY   *(integer multiply; currently causes a trap)*
DIV0   *(initialize divide)*
DIV   *(divide step)*
DIVL   *(divide last step)*
DIVREM   *(divide remainder)*
DIVIDE   *(integer divide; currently causes a trap)*

Logical Operations

AND, OR, XOR
ANDN   *(AND not)*
NAND   *(not AND)*
NOR   *(not OR)*
XNOR   *(not XOR)*

SLL   *(shift left logical)*
SRL   *(shift right logical)*
SRA   *(shift right arithmetic)*

CLZ   *(count leading zeroes)*

(continued)

Table 10. (continued)

Field Extraction and Insertion Operations

EXBYTE	*(extract byte)*
EXHW	*(extract half-word)*
EXHWS	*(extract half-word, sign-extended)*
EXTRACT	*(bit aligned word extract from two words)*
INBYTE	*(insert byte)*
INHW	*(insert half-word)*
CONST	*(move 16-bit immediate to register, zero/ones extended)*
CONSTH	*(replace register high half-word with 16-bit immediate value)*

Loads and Store Instructions

LOAD	*(load from memory or coprocessor)*
LOADL	*(load from memory or coprocessor, assert LOCK)*
LOADM	*(load multiple contiguous registers)*
LOADSET	*(load from memory and write ones back to the same location, assert LOCK)*
STORE	*(store to memory or coprocessor)*
STOREL	*(store to memory or coprocessor, assert LOCK)*
STOREM	*(store multiple contiguous registers)*

Special Register Instructions

MFSR/MTSR	*(move from/to special register to/from general register)*
MTSRIM	*(move 16-bit immediate to special register)*
MFTLB/MTTLB	*(move from/to TLB register to/from general register)*
SETIP	*(set register indirection pointers)*

(continued)

Table 10. (continued)

Comparison Operations
The compare instructions set the destination to TRUE if the sources
(one register and a second register or 8-bit immediate value)
meet the condition; otherwise it is set to FALSE.

CPEQ	*(compare equal)*
CPNEQ	*(compare not equal)*
CPB	*(compare bytes; TRUE if at least one byte matches)*
CPGE	*(compare greater than or equal, signed or unsigned)*
CPGT	*(compare greater than, signed or unsigned)*
CPLE	*(compare less than or equal, signed or unsigned)*
CPLT	*(compare less than, signed or unsigned)*
ASEQ/ASNEQ	*(assert equal/not equal to; trap if FALSE)*
ASGE	*(assert greater than or equal to, signed or unsigned; else trap)*
ASGT	*(assert greater than, signed or unsigned; else trap)*
ASLE	*(assert less than or equal to, signed or unsigned; else trap)*
ASLT	*(assert less than, signed or unsigned; else trap)*

Program Flow Control Instructions

JMP	*(jump, 16-bit absolute or PC relative)*
JMPI	*(jump to address specified in Rb)*
JMPTI	*(jump if register Ra TRUE, 16-bit absolute or PC relative)*
JMPT	*(jump if register Ra TRUE to address specified in Rb)*
JMPFI	*(jump if register Ra FALSE, 16-bit absolute or PC relative)*
JMPF	*(jump if register Ra FALSE to address specified in Rb)*
JMPFDEC	*(jump if register Ra FALSE, 16-bit absolute or PC relative, decrement Ra)*

(continued)

Table 10. (continued)

Subroutine calls
(the return address is copied to a specified register)

CALLI	*(call subroutine, 16-bit absolute or PC relative)*
CALL	*(call subroutine at address specified in register Rb)*
IRET	*(interrupt return)*
IRETINV	*(interrupt return and invalidate branch target cache)*
EMULATE	*(trap to software emulation)*

Processor Control Instructions

HALT	
INV	*(invalidate branch target cache)*

Floating-Point Operations (currently cause a trap)

CVDF	*(convert double- to single-precision)*
CVDINT	*(convert double-precision to integer)*
CVFINT	*(convert single-precision to integer)*
CVFD	*(convert single- to double-precision)*
CVINTD	*(convert integer to double-precision)*
CVINTF	*(convert integer to single-precision)*
DADD/DSUB	*(double-precision add/subtract)*
DDIV	*(double-precision divide)*
DMUL	*(double-precision multiply)*
FADD/FSUB	*(single-precision add/subtract)*
FDIV	*(single-precision divide)*
FMUL	*(single-precision multiply)*

(continued)

Table 10. (continued)

---

*The following compare instructions set the destination to TRUE
if the sources (both floating-point values) meet the condition;
otherwise it is set to FALSE:*

DEQ	*(double-precision equal to)*
DGT	*(double-precision greater than)*
DLT	*(double-precision less than)*
FEQ	*(single-precision equal to)*
FGT	*(single-precision greater than)*
FLT	*(single-precision less than)*

(Source: AMD, 1987)

3.6.2 Organization

The top-level organization of the chip is shown in figure 68. It is a single address bus Harvard architecture. Though an instruction is used every cycle, most instruction fetches are sequential, so a burst-mode memory does not require new address information for each fetch. The address bus can therefore be used to direct data transfers much of the time without impacting instruction fetches.

The chip has an instruction prefetch mechanism which allows instructions to be loaded before the execution unit has requested them. This is performed by a *prefetch address generator*, which increments the instruction address from the point of the last requested address, and the *instruction prefetch buffer*, which holds the prefetched instructions until the execution unit is ready for them. The prefetch address generator operates with physical addresses, and is limited to incrementing within one kilobyte aligned blocks of memory, so that it can never increment across a virtual page boundary. When the boundary is reached, prefetching ceases until the execution unit makes a request which causes an address translation into the next physical page. This address translation includes the necessary protection violation checks.

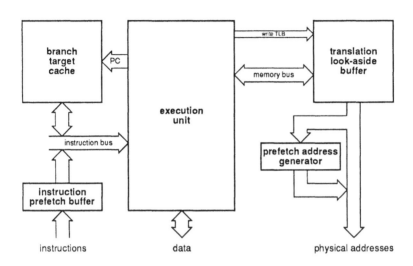

Figure 68: Am29000 chip block diagram
(Adapted from AMD, 1987)

Whenever a non-sequential instruction address is generated, the *branch target cache* copies the first four words to be fetched from the new address. The next time a branch goes to the same destination, the branch target cache will indicate a hit and supply the first four instructions from that address. The processor will request instructions from external memory starting four words on from the branch destination. The external memory is likely to have a much longer non-sequential access time than its normal sequential access time, so the branch target cache delivers the instructions which are needed to keep the execution unit busy during this latency period. Figure 69 illustrates the ideal case, where the instruction memory is assumed to have a one cycle access for sequential addresses and a five cycle access for non-sequential addresses.

The branch target cache is a dual-set associative cache with random replacement (based on the processor clock). Each set contains 16 lines of four words, and the lines can start from any memory word. Normally, an instruction cache with a quad-word line size would store quad-word aligned blocks, but with this special form of cache such a constraint would be too restrictive. Each line has an associated address tag, and a process ID tag.

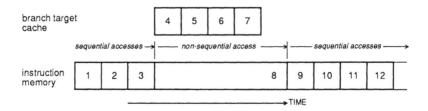

Figure 69: Am29000 branch target cache operation

Normally all four words are valid or invalid together, but a second taken branch within the block, or a page boundary crossing, may cause some words to remain invalid. If the corresponding instructions are subsequently requested, a separate instruction fetch will be initiated.

The *translation look-aside buffer* (TLB) is also a dual-set associative cache, this time with a least recently used replacement algorithm. Each set contains 32 translation entries, with a virtual address tag, process ID bits, and access protection information. The page size may be configured to 1Kbytes, 2Kbytes, 4Kbytes or 8Kbytes. When a TLB miss occurs, an exception is generated, and system software is responsible for updating the TLB appropriately. The only hardware assistance available is a least recently used register which directs the software to the appropriate TLB entry to put the new page table entry into.

The *execution unit* is shown in more detail in figure 70. The *PC unit* is completely separate from the general-purpose registers and datapath, and contains the recent PC history which is required to recover from exceptions in a processor which supports delayed branches. The PC unit increments the program counter, but requires assistance to handle branches. It supplies the appropriate PC as a subroutine return address to the register file, and the old PC values may be inspected and modified by transfer to and from the datapath.

The *address unit* contains a 30-bit address adder which is used to calculate both instruction and data addresses. Branches use the current PC plus an offset specified in the instruction. Single register loads and stores use register addressing, and do not require additional address calculations, so these bypass the address unit. Load and store multiple register instructions use register addressing for the first transfer, but then the address must be incremented

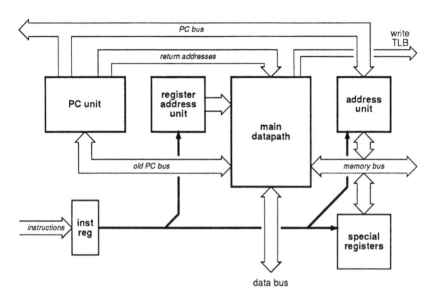

Figure 70: The Am29000 execution unit
(Adapted from AMD, 1987)

until all transfers are complete. The address unit does the incrementing, and retains the incremented value in case a burst-mode transfer requires restarting. The address unit is also responsible for extracting immediate data values from instructions.

The *special registers* contain status and configuration information, much of which is protected from user access. They include the channel control registers, which contain the information which is needed to retry a data transfer which has caused a page fault, the LRU recommendation for updating the TLB, and so on. These registers must be copied to and from general-purpose registers if they are to be changed; otherwise they are only modified by the side-effects of operations on general-purpose registers.

The *register address generator* calculates the operand and destination addresses in the register bank. Since register addressing can be stack pointer relative, there is an 8-bit adder on each of the three address generators. The register address incrementer for load and store multiple instructions is here also, as is a write address register for load data. All register and data race hazards are detected here.

Since the register specifiers are in fixed fields in the instructions, register addresses can be calculated in parallel with instruction decode, and the operands fetched. This overlap reduces the effective pipeline depth.

The *main datapath* (figure 71) contains the general-purpose registers and the function units. The three function units are an ALU, a priority encoder, and a field shift unit which performs general shifts and byte and half-word extracts and inserts. The priority encoder outputs the number of leading zeroes in the input word. The ALU supports arithmetic, bitwise logical and relational operations, and includes support for multiply and divide steps.

The datapath operates in three pipeline stages. Operand fetch is concurrent with instruction decode in the first stage. Execution in one of the function units takes the second stage, and writing the result is the third. A register which is written in the same cycle that it is read will produce the written value, so there are no data hazards there. Register forwarding is used to avoid hazards between the execute and read/write pipeline stages.

The Am29027 Floating-Point Accelerator

The Am29000 architecture allows for the connection of coprocessors, and the Am29027 floating-point accelerator is one such coprocessor. The Am29027 supports a number of floating-point formats, including the 32- and 64-bit IEEE standards. The Am29000 can send 64 bits of information to the Am29027 along the address bus and data bus in parallel for high bandwidth communication.

The Am29027 may be configured to optimize either throughput or latency. Vectorized calculations will benefit from use of the pipelined mode, which maximizes the throughput, whereas a single calculation will be performed faster in flow-through mode.

An Am29000 System

A block diagram of a simplified Am29000 system is shown in figure 72. The instruction memory may be constructed from video RAM (VRAM); this can supply 25 million sequential instructions per second, which is the peak requirement of the CPU. Non-sequential instructions are much slower, but the branch target cache is designed to cover for these.

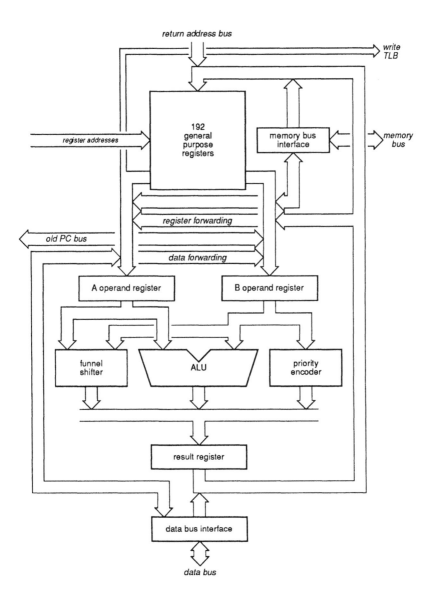

Figure 71: Am29000 main datapath
(Adapted from AMD, 1987)

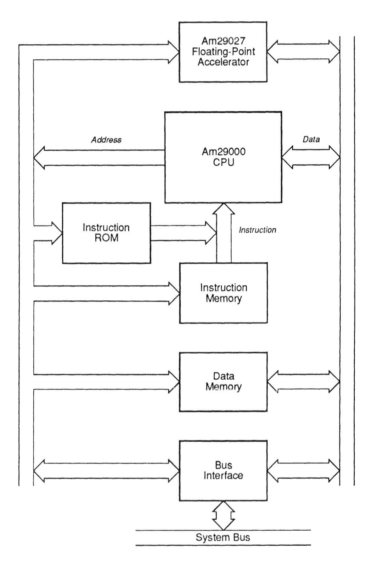

Figure 72: Am29000 system block diagram
(From AMD, 1987)

3.6.3 Conclusions

The Am29000 has a number of features which differentiate it from other RISC processors. The principle ones are:

(1)   A very large, uniformly-addressable register bank.

The 192 general-purpose registers are not unusually large for a RISC CPU, but the fact that they are all visible at once is. The Berkeley register windows scheme has been abandoned in favour of register stack pointer relative addressing, which allows the system to select the stack frame size rather than having it frozen in hardware. The negative side of this feature is that every register specifier field in an instruction uses 8 bits, as against the more usual 5. Since most instructions have three such fields, 9 extra bits are required to specify the registers. This has a cost in terms of code expansion and instruction bandwidth.

(2)   A branch target cache.

It is a characteristic of most memory systems that they can supply data from sequential locations much faster than they can from random addresses. The branch target cache is a very effective application of a small cache towards filling in the gaps left by the slower random accesses. When semiconductor technology advances to the point where all CPUs have large on-chip caches this specialized sort of cache will give way to conventional caches, which give the same benefits and more. While the chip area only allows relatively small caches, this sort of architecture may well be the best use of the space.

(3)   A large number of special-purpose registers.

The Am29000 supports a number of complex instructions, but does them in fairly simple ways. Intermediate state is kept in special registers, where it may be accessed to enable the instruction to be resumed should an exception arise during execution. For instance, a load multiple registers could cross a page boundary and cause a page fault part way through. The page fault may result in a page swap, and while waiting for the page to be brought in, the scheduler may wish to activate another process. The state of the load multiple may be saved by copying the *load/store count remaining* special register, and restoring it before resuming the faulted task. Managing all the special registers is a complex task for the system software.

(4)   The delayed load mechanism.

The register bank has a single write port, so when a loaded data item arrives it may not be possible to copy it into the destination register. The data is held in a latch until the write port is free. This may take an arbitrary time, and in the mean time instructions may operate on the fetched data, so an automatic forwarding mechanism bypasses the register bank when this happens. The only limit to the time the data may wait before being written to the destination is that any subsequent load instruction will not use its write slot, so there can never be more than one data item outstanding at any time.

(5)   It is suggested that video RAM can be used to supply the very high bandwidth instruction stream required by the Am29000. This may be a very cost effective solution compared with an instruction cache built from static memory. It is also a good match to the characteristics of the branch target cache.

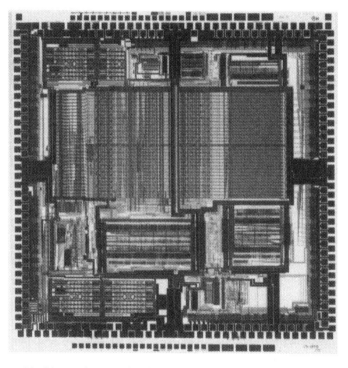

Figure 73: Photomicrograph of an NMOS implementation of the HPPA
(©Copyright 1986 Hewlett-Packard Company.
Reproduced with permission.)

3.7 THE HP PRECISION ARCHITECTURE

The Hewlett-Packard Precision Architecture (HPPA) is the outcome of a massive development program to produce a scalable system architecture which is capable of covering HP's computer and instrumentation markets.

From the outset, the aim was to allow for many different implementations with different cost/performance levels but with a common instruction set. Extensive studies were made of instruction set usage, not only on the basis of laboratory benchmark programs, but also on the workloads of many customers using HP's existing three computer lines.

The overall approach resembles the approach of the Berkeley RISC team, but the scale of the work and the resources applied to it are of a different order of magnitude. The HPPA is the RISC approach applied on a grand scale.

See Hewlett-Packard (1986) for complete details of the Precision Architecture.

3.7.1 System Organization

Before we examine the processor architecture in detail, it is worth looking at the overall system organization. The hardware structures are defined hierarchically, with the CPU details at a low level in the hierarchy (and differing between one implementation and another).

At the highest level in the hardware hierarchy is the overall system structure (figure 74). The high-performance elements are clustered around a central bus, which will support processor, memory and I/O modules in various configurations. High-speed I/O will be connected to this bus, but low speed I/O will be connected to remote buses through bus adapter modules. The central bus uses 32-bit physical addresses, and all memory and I/O use the same addressing and protection mechanisms.

At the next level in the hierarchy is the organization of the processor node (figure 75). The CPU interfaces to the central bus through data and instruction caches (usually), and the node contains an address translation mechanism when virtual memory is to be supported. Three levels of system are defined. HP Precision level 0 does not use virtual addressing, level 1 uses 48-bit virtual addressing (in the form of 2^{16} 32-bit virtual spaces), and level 2 uses 64-bit virtual addressing (in the form of 2^{32} 32-bit virtual spaces). Higher levels are functional supersets of lower levels.

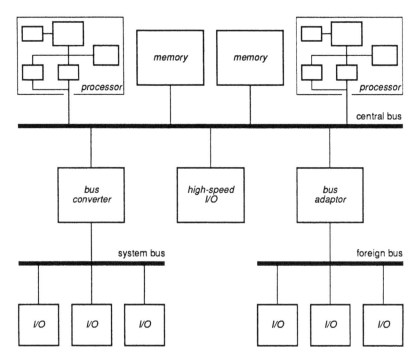

Figure 74: HP precision architecture system organization
(From Hewlett-Packard, 1986)

The functionality of the CPU may be extended in two ways. *Coprocessors* may be added; these typically have their own register sets, and perform independent calculations concurrently with other CPU activity. *Special Function Units* are logical extensions to the on-chip function units and operate on values in the CPU registers, for instance to enhance signal processing capability.

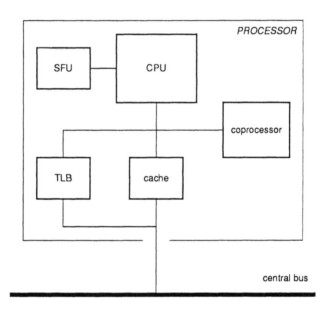

Figure 75: HP precision architecture processor organization
(From Hewlett-Packard, 1986)

3.7.2 Architecture

The HP Precision Architecture defines the programmer's model, the supported data formats, and the instruction set of all implementations.

The data formats are:

- Signed and Unsigned bytes, halfwords and words. Halfwords must be halfword aligned; words must be word aligned.

- Single (32-bit), double (64-bit) and extended (128-bit) floating-point numbers. Single-precision numbers must be word-aligned, double- and extended-precision numbers must be aligned on doubleword boundaries.

- Packed and unpacked decimal numbers. Packed decimal numbers are aligned to word boundaries, and consist of 7, 15, 23 or 31 BCD digits (each four bits long) plus a 4-bit sign.

The processor's state is held in a large number of registers (figure 76). There are 32 general-purpose registers, which is not unusual, but there are also a large number of special-purpose control registers. Many of these are for operating system use only. In part the unusually large number is explained by the very large virtual address space supported by the architecture; all the *space* registers are preserving virtual address extensions for various purposes. (This includes the IASQ registers and the space registers in the control register bank, as well as the eight registers in the space register bank.)

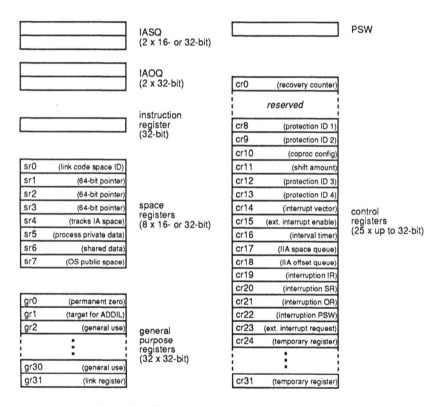

Figure 76: HP precision architecture register set
(From Hewlett-Packard, 1986)

The twelve formats used by the instruction set are shown in figure 77. Though this is a high number by RISC standards, it can be seen from the figure that there is a great deal of commonality between the various formats.

In particular, the register specifier fields are in fixed positions. This allows operand access to be performed in parallel with instruction decoding.

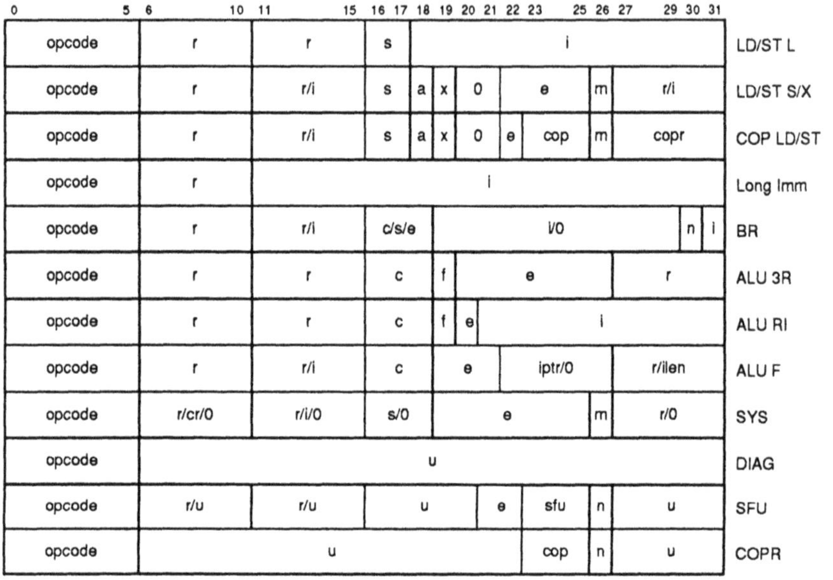

0	5 6	10 11	15	16 17	18	19	20 21 22 23		25 26 27		29 30 31	
opcode	r	r	s				i					LD/ST L
opcode	r	r/i	s	a	x	0	e		m	r/i		LD/ST S/X
opcode	r	r/i	s	a	x	0	e	cop	m	copr		COP LD/ST
opcode	r				i							Long Imm
opcode	r	r/i	c/s/e			i/0				n	i	BR
opcode	r	r	c	f		e				r		ALU 3R
opcode	r	r	c	f	e		i					ALU RI
opcode	r	r/i	c		e	iptr/0			r/ilen			ALU F
opcode	r/cr/0	r/i/0	s/0		e			m	r/0			SYS
opcode			u									DIAG
opcode	r/u	r/u	u		e	sfu		n	u			SFU
opcode		u				cop		n	u			COPR

r	: general register specifier	m	: modification specifier	cr	: control register
s	: space register specifier	n	: nullification specifier	0	: set to zeros
i	: immediate	c	: condition specifier	u	: undefined (instruction extension)
a	: pre/post, or shift index	f	: falsify condition c	sfu	: special function unit identifier
x	: indexed/short displacement	iptr	: immediate pointer	cop	: coprocessor unit identifier
e	: opcode extension	ilen	: immediate length	copr	: coprocessor register

Figure 77: HP precision architecture instruction formats
(From Mahon et al, 1986)

An unusual feature of the instruction set is the use of fragmented immediate fields. Whenever an instruction uses an immediate value, all the bits which are not required for a specific purpose are used to construct a maximum length immediate value. The branch instructions, for example, may have three separate fields which are concatenated to form the immediate offset. The hardware described in the next section has a block called the *immediate field assembler* to put them back together!

The instruction set is summarized in table 11. Note that:

- Loads have delayed action, but hardware interlocks control race hazards. A compiler should attempt to put instructions after the load that do not use the loaded value, but the processor will operate correctly (though more slowly) if this is not done.

- The load and store addressing modes are relatively complex. Stores cannot use base plus index addressing, presumably to allow single cycle execution with a register bank with only two read buses.

- The extensive use of conditional skips is unusual. It requires a bit in the PSW to latch the nullify output of the computation instruction in order to disable the following instruction. The nullify bit must automatically clear itself after one instruction.

3.7.3 Organization

The internal organization of one NMOS implementation of a Precision Architecture CPU is shown in figure 78. (See Forsyth et al, 1987 for further details.)

This chip implements the entire HPPA instruction set using direct hardwired decoding. It is built on a 1.5 micron double metal NMOS process, and uses 115,000 transistors on a 8.4mm x 8.4mm die. With a 30MHz clock it processes 15 million instructions per second (peak), and consumes about 10 watts of power.

The pipeline is broken down into five stages:

(1)   Instruction fetch.

(2)   Instruction decode/operand fetch.

(3)   Execute/ALU operation.

(4)   Data access.

(5)   Result write.

Each stage takes one clock period, and a new instruction starts every second clock period. By defining a machine cycle to be two clock periods, the designers claim that 93% of the instructions take one cycle.

There is a single 32-bit cache interface, and it is used alternately for the data access of one instruction and then the instruction fetch of the next instruction but one. A branch instruction uses its data access for fetching the branch target, so a conditional branch has both the sequential instruction

Table 11. The HP precision architecture instruction set

Memory Reference Instructions

LD	*(byte, halfword, word,*
	base plus 14-bit immediate offset,
	with optional post-increment or pre-decrement,
	or base plus 5-bit immediate offset,
	with optional pre- or post- increment or decrement,
	or base plus index (optionally scaled to data size),
	with optional pre-increment)
LDCWX	*(load and clear word, indexed or 5-bit offset)*
ST	*(byte, halfword, word,*
	base plus 14-bit immediate offset,
	with optional post-increment or pre-decrement,
	or base plus 5-bit immediate offset,
	with optional pre- or post- increment or decrement)
STBYS	*(store mis-aligned bytes, 5-bit offset)*

Branch Instructions
(branches have a single delay slot which may optionally be nullified)

BL	*(branch PC relative and copy return address to register,*
	using 17-bit immediate or register offset)
GATEWAY	*(branch 17-bit PC relative and change privilege level)*
BV	*(branch register relative with scaled register offset)*
BE	*(branch external, ie to another space; optionally with link)*
MOVB	*(move and branch conditionally, 12-bit PC relative)*
COMB	*(compare and branch conditionally, 12-bit PC relative)*
ADDB	*(add and branch conditionally, 12-bit PC relative)*
BB	*(branch on bit, 12-bit PC relative)*

(continued)

Table 11. (continued)

Computational Instructions

ADD	*(add and conditionally skip the next instruction,*
	optional trap on overflow, optionally set PSW condition codes)
ADC	*(add with carry and conditionally skip the next instruction,*
	optional trap on overflow)
SHnADD	*(shift 1, 2 or 3 bits and add, conditionally skip*
	the next instruction, optional trap on overflow)
SUB	*(subtract, with or without borrow,*
	conditionally skip the next instruction,
	optional trap on condition and/or overflow)
DS	*(divide step)*
COMCLR	*(compare 2 registers, clear a third,*
	conditionally skip the next instruction)
OR	*(bitwise logical OR, conditionally skip the next instruction)*
XOR	*(bitwise logical XOR, conditionally skip the next instruction)*
AND	*(bitwise logical AND, conditionally skip the next instruction)*
ANDCM	*(bit clear and conditionally skip the next instruction)*
UXOR	*(compare corresponding sub-units for equality,*
	conditionally skip the next instruction)
UADDCM	*(compare corresponding sub-units for inequality,*
	conditionally skip the next instruction or trap)
DCOR	*(to decimal correct the eight BCD digits in a register,*
	conditionally skip the next instruction)
SHD	*(shift of two concatenated registers,*
	conditionally skip the next instruction)
EXT	*(extract bit field and conditionally skip the next instruction)*
DEP	*(deposit bit field and conditionally skip the next instruction)*

(continued)

Table 11. (continued)

Long Immediate Instructions

LDO	*(load effective address into register)*
LDIL	*(load 21-bit immediate into most significant end of register)*
ADDIL	*(add 21-bit immediate into most significant end of register)*

System Control Instructions

BREAK	*(break to debugger)*
RFI	*(return from interrupt)*
SSM	*(set system mask)*
RSM	*(reset system mask)*
MTSM	*(move to system mask)*
LDSID	*(load space identifier)*
MT/FSP	*(move to/from space register)*
MT/FCTL	*(move to/from control register)*
SYNC	*(synchronize caches)*
PROBER/W	*(probe read/write access to see if address is allowed)*
LPA	*(load physical address)*
LHA	*(load hash address to locate hash table entry)*
PD/ITLB	*(purge data/instruction TLB)*
PD/ITLBE	*(purge data/instruction TLB entry)*
ID/ITLBA/P	*(insert data/instruction TLB address/protection)*
PDC	*(purge data cache)*
FD/IC	*(flush data/instruction cache)*
FD/ICE	*(flush data/instruction cache entry)*
DIAGNOSE	*(trap to diagnostic or emulation software)*

(continued)

Table 11. (continued)

Assist Instructions

SPOP0	*(invoke a special function unit operation,*
	conditionally skip the next instruction)
SPOP1	*(copy a special function unit register to a general register,*
	conditionally skip the next instruction)
SPOP2	*(invoke a parameterized special function unit operation,*
	conditionally skip the next instruction)
SPOP3	*(invoke a special function unit operation with two parameters,*
	conditionally skip the next instruction)
COPR	*(coprocessor operation, conditionally skip the next instruction)*
CLD/ST	*(coprocessor load/store, word or doubleword,*
	the addressing is similar to LD/ST)

(Source: adapted from Hewlett-Packard, 1986)

(automatically prefetched by the cache) and the target to choose between when it has the result of the condition evaluation.

3.7.4 Conclusions

The Hewlett-Packard Precision Architecture is an architectural specification which uses many RISC ideas, but goes a lot further than just defining an instruction set for efficient hardware implementation. Some aspects of the system suggest that considerable emphasis has gone into giving the architecture a lot of upward growth potential, and this has introduced considerable complexity into the specification.

Most of this complexity is associated with managing the very large virtual address space that the specification demands. It is a repeated error of computer architects that they allow insufficient virtual address bits in a new architecture; applications use up the available virtual address space well before the machine is obsolete. The Hewlett-Packard Corporation clearly wishes to avoid this pitfall with the Precision Architecture.

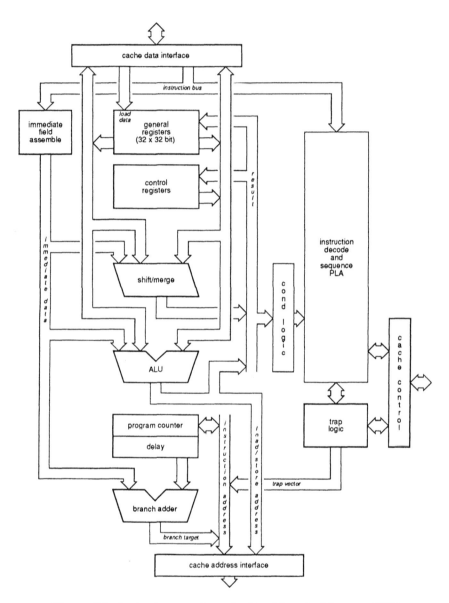

Figure 78: An implementation of a Precision Architecture CPU
(From Forsyth et al, 1987. ©1987 IEEE.)

The most unusual detail of the architecture is the option in many instructions to conditionally skip the next instruction. This requires a transient bit of processor state which must be preserved and restored across exceptions and process swaps. Computer designers usually take pains to minimize such features as they make an implementation harder to validate.

Despite the relative complexity of the architectural requirements, the entire CPU has been built with 115,000 transistors, which is not a large number by present VLSI standards.

3.8 THE MOTOROLA M88000

The Motorola M88000 architecture combines a RISC-like instruction set with multiple concurrent pipelined function units modelled on the CDC 7600 supercomputer organization (Dobbs, Reed and Ng, 1988). It is the first RISC product from Motorola Inc., who are a dominant force in the 32-bit CISC market with their 68000 family, and its market positioning is therefore of considerable interest.

Unlike the Intel Corporation, another company with a strong 32-bit CISC market position, Motorola has positioned its RISC product in a similar market to its established CISC products. At the same time, it has announced continued commitment to the 68000 series with further performance-enhanced versions being developed. (See section 3.9 for further discussion of Intel's positioning.)

The MC88100 is the first processor in the M88000 family. It is characterized by a modified Harvard architecture with a reduced instruction set, high performance floating-point on chip, and multiple function units which allow up to 11 instructions to be in process in one clock cycle. It is supported by the MC88200 cache and memory management unit (CMMU), which has a 16Kbyte four-way set associative cache and an address translation unit on chip. One MC88100 requires at least two MC88200 devices (one for instructions and one for data), but configurations with more of either chip allow multiprocessing, larger caches and fault detection.

The CPU and CMMU are both built on a 1.5 micron CMOS process. The MC88100 uses 165,000 transistors, the MC88200 uses 750,000 transistors. With a 20MHz clock the quoted sustainable performance is 15 to 17 MIPS and 7 MFLOPS.

3.8.1 Architecture

The 88100 processor supports the following data types:

(1)   Signed and unsigned bytes, halfwords and words.

(2)   Single and double precision IEEE standard floating-point numbers.

(3)   Signed and unsigned bit fields from 1 to 32 bits long, specified by a five-bit width field and a five-bit offset within a word.

The user has access to a bank of 32 general-purpose 32-bit registers (figure 79), which are used for all operands. Double precision floating-point operands occupy two adjacent general registers. All data dependencies are handled by

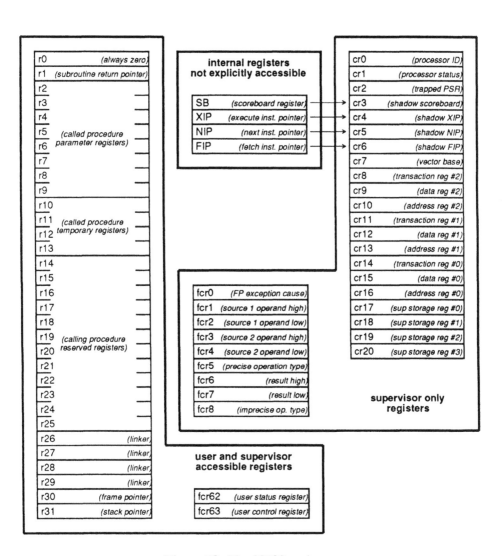

Figure 79: The 88100 registers
(From Motorola, 1988)

hardware interlocks. The supervisor has a considerably larger register set, mainly concerned with supplying information for exception recovery.

The various instruction formats are shown in figure 80. There is a high degree of regularity in the positions of the fields, and in particular the source register fields are in fixed positions to allow operand fetch to proceed concurrently with instruction decode. The 88100 has a load/store architecture. The instruction set is listed in table 12.

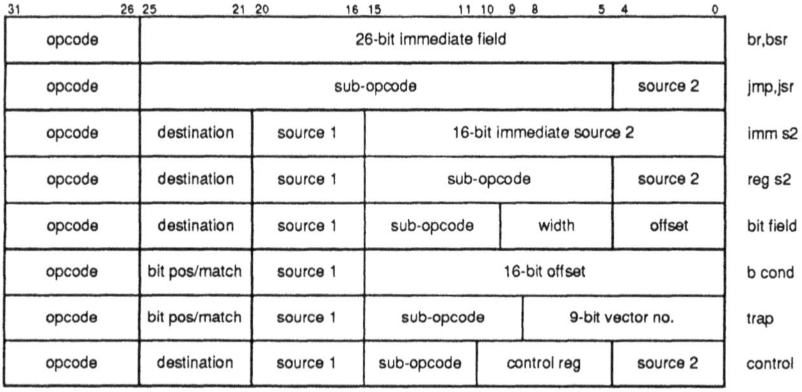

31	26 25	21 20	16 15	11 10 9 8	5 4	0	
opcode	26-bit immediate field						br,bsr
opcode	sub-opcode				source 2		jmp,jsr
opcode	destination	source 1	16-bit immediate source 2				imm s2
opcode	destination	source 1	sub-opcode		source 2		reg s2
opcode	destination	source 1	sub-opcode	width	offset		bit field
opcode	bit pos/match	source 1	16-bit offset				b cond
opcode	bit pos/match	source 1	sub-opcode	9-bit vector no.			trap
opcode	destination	source 1	sub-opcode	control reg	source 2		control

Figure 80: 88100 instruction formats

The most notable feature of the instruction set is the extensive support for bit-field operations and tests. Most RISC architectures have good shifting instructions, but explicit bit-field types are not usually supported.

Store instructions with base plus index addressing are supported.

3.8.2 Organization

A minimum processor node is illustrated in figure 81. The internal organization of the processor is based around three operand buses, for the destination and two sources of a typical operation. The architecture allows for the addition of six *special function units*, which can extend the instruction set in a symetrical way. The integer and floating-point units are special function units 0 and 1 respectively.

Table 12. The MC88100 instruction set

Integer Arithmetic Instructions
(second operand is a register or 16-bit immediate field)

add	*(add signed or unsigned, optional carries in/out)*
sub	*(subtract signed or unsigned, optional borrows in/out)*
div	*(divide signed or unsigned)*
cmp	*(compare)*
mul	*(multiply)*

Logical Instructions
(second operand is a true or complemented register,
or a 16-bit immediate field aligned to upper or lower halfword)

and	
or	
xor	
mask	*(logical mask, 16-bit immediate upper or lower halfword only)*

Bit-Field Instructions
(width and offset may be specified in a register or an immediate field)

clr	*(clear bit field)*
set	*(set bit field)*
mak	*(make bit field)*
ext	*(extract signed or unsigned bit field)*
ff	*(find first bit, clear or set)*
rot	*(rotate register)*

(continued)

Table 12. (continued)

Memory Reference Instructions
(base plus 16-bit immediate offset addressing,
or base plus index optionally scaled to data size)

ld	*(load register from memory)*
st	*(store register to memory)*
xmem	*(exchange register with memory)*
lda	*(load address)*

Control Register Instructions

ldcr	*(load from control register)*
stcr	*(store to control register)*
xcr	*(exchange register with control register)*

Flow Control Instructions
(branches and jumps optionally execute the following instruction)

bb	*(branch on bit clear or set, 16-bit PC relative)*
bcnd	*(conditional branch, 16-bit PC relative)*
br	*(unconditional branch, 26-bit PC relative)*
bsr	*(branch to subroutine, 26-bit PC relative)*
jmp	*(unconditional jump to address in register)*
jsr	*(jump to subroutine at address in register)*
rte	*(return from exception)*
tb	*(trap on bit clear or set to 9-bit trap vector)*
tbnd	*(trap on bounds check to implicit trap vector)*
tcnd	*(conditional trap to 9-bit trap vector)*

(continued)

Table 12. (continued)

Floating-Point Arithmetic Instructions

fadd	*(floating-point add)*
fdiv	*(floating-point divide)*
fmul	*(floating-point multiply)*
fcmp	*(floating-point compare)*
flt	*(convert integer to floating-point)*
int	*(round floating-point to integer)*
nint	*(round floating-point to nearest integer)*
trnc	*(truncate floating-point to integer)*
fldcr	*(load from floating-point control register)*
fstcr	*(store to floating-point control register)*
fxcr	*(exchange floating-point control register)*

(Source: Motorola, 1988)

The register file has an associated *sequencer*, which is responsible for managing data dependencies. During instruction decode, the source operands are fetched from the register bank, and the scoreboard bits corresponding to those registers are also accessed. The scoreboard bit for the destination is accessed at the same time. If any of the scoreboard bits shows that a source or destination register is locked, the instruction cannot begin. When an instruction begins it locks its destination register, and when the result is written the lock is cleared. Several instructions can be in process if there is no contention for any of their destination registers.

The floating-point unit contains a 32x32 combinatorial multiplier, which is used for integer multiplication as well as floating-point. The combinatorial multiplier allows the FPU to complete a single-precision multiply every cycle, through a six-cycle pipeline. The floating-point adder can likewise complete an addition every cycle, through a five-cycle pipeline.

The MC88100 uses a modified Harvard architecture, and has separate instruction and data ports. These ports both use the same bus protocol, called the P bus. This bus is capable of delivering a word of read data every cycle,

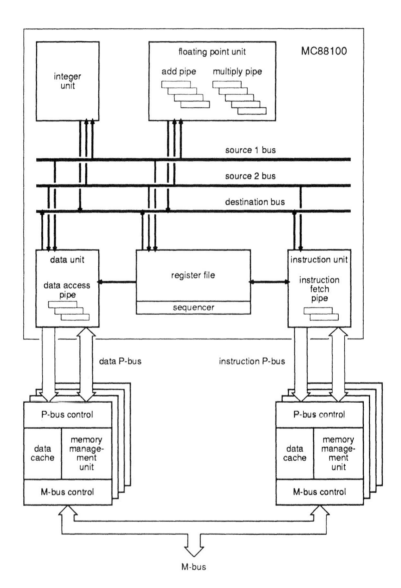

Figure 81: The 88000 system organization
(From Motorola, 1988)

with an address to data delay of less than a cycle. Handshake information is pipelined. At the other end of each P bus there will normally be one or more of the MC88200 CMMU chips. These chips have an M bus on the other side, which is a bus for connecting memory systems and I/O.

The MC88200 CMMU

The MC88200 combines the functions of memory management and data/instruction caching. It can translate addresses by one of two mechanisms:

(1)   The *block address translation cache* is an associative translation buffer for ten 512Kbyte blocks of memory. Eight of the entries are maintained by software, and the remaining two are hardwired for memory-mapped I/O.

(2)   The *page address translation cache* is a fully associative 56-entry translation buffer with automatic update. Each entry contains the logical to physical translation for one 4Kbyte page, and a translation miss causes the P bus to give a 'wait' reply while the translation is located in a two-level page table in main memory.

The cache is four-way associative with physical address tags. It uses a least recently used algorithm to select between the sets. Cache look-up proceeds in parallel with address translation, and the physical page number is compared with the tag from each of the four cache sets in parallel. If a match is found, the corresponding cache data is accessed.

The cache has a quad-word line length, and cache loads always access complete lines from quad-word aligned addresses. Each line has a valid, disable, and modified bit. The disable bit is used by software to prevent the system from using a cache line which is malfunctioning; there are facilities for testing individual cache lines, and this provides a measure of fault tolerance in the cache memory.

The memory update strategy is determined by bits in the individual page table entries. Write-through operation ensures memory consistency at all times, whereas copy-back operation minimizes bus traffic. The copy-back mechanism involves sending the first write to a particular location out to main memory to ensure that no errors arise, but thereafter writes to that location only update the cache.

The MC88200 supports a cache coherency protocol to ensure that multiprocessors have a consistent view of memory. All M bus transactions are monitored, and when another device attempts to access a location which has been updated in the CMMU's cache but not copied back, the M bus transaction is preempted by the monitoring MC88200 so that it can correct any inconsistencies before the transaction resumes.

3.8.3 Conclusions

The M88000 family is a conventional RISC architecture (instruction set, register set) in a sophisticated system organization. The concurrent function units allow considerable parallelism in execution, and the scoreboard mechanism allows the functionality to be extended in the future.

The CMMU chip contains a large cache and a conventional address translation mechanism. It converts the extremely demanding P bus into the more manageable M bus, and the M bus provides directly for multiprocessing.

The M88000 offers a set of components well suited to high-performance uniprocessor or multiprocessor applications.

3.9 THE INTEL 80960KB

The Intel 80960KB is the first member of a family of 32-bit RISC processors intended for embedded controller applications. The Intel Corporation is one of the most powerful forces in the microprocessor industry, and has built this position on the basis of very complex microcoded 8-, 16- and 32-bit CPUs. The 80960 is their first product based on RISC architecture, and both the design and the market positioning are of great interest.

In targetting the 80960 family at embedded applications, Intel are attempting to avoid confusion in the market where their existing 32-bit product (the 80386) and its successors (the 80486,...) have established a firm hold. These chips are dominant in the high-end PC and low-end engineering workstation areas. The usual RISC target market is in the high-end engineering workstation, minicomputer and server areas, but Intel is steering the 80960 well away from all general-purpose applications. It is interesting to contrast this approach with that of Motorola, the other major player in the 32-bit CISC microprocessor business. They (Motorola) are positioning their RISC product much closer to their existing 68020/68030 market.

The only technical feature of the 80960 which precludes its use as a general-purpose workstation CPU is the absence of any memory management. If, in the future, the 80386 family were to begin to lose ground to competition from RISCs, it might not take Intel too long to add an MMU and reposition the 80960.

The 80960KB is built on a CMOS process, and with a 20MHz clock it has a sustainable performance of 7.5 MIPS.

3.9.1 Architecture

The architecture defines data types of 8-, 16-, and 32-bit signed and unsigned integers, 32-, 64-, and 80-bit IEEE standard real numbers, and bits, bit fields, triple-words and quad-words.

The programmer visible state is contained in 40 registers (figure 82). Four are special-purpose control registers, and four are for floating-point values. Any general register may be used for floating-point values, but the the special registers are 80 bits long and make the handling of extended precision more efficient. The remaining 32 registers are general-purpose, though some have special functions to do with stack management.

Of the general-purpose registers, 16 are global, and 16 are local to the current procedure. There is an on-chip register cache which contains four

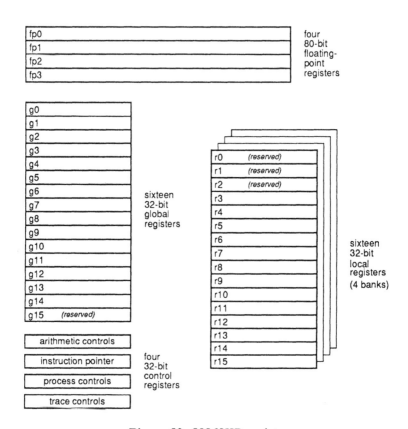

Figure 82: 80960KB registers

banks of 16 local registers, and procedure entry and exit cause the bank to switch. If the procedure calls nest beyond a depth of four, the CPU automatically copies a bank onto a stack in main memory to avoid underflow or overflow. Future implementations may have more register banks in a programmer transparent way.

The programmer sees a 4 Gbyte linear address space, with no address translation. All memory references are by LOAD and STORE instructions.

There are five instruction formats (figure 83), all taking 32 bits (though the long memory reference has an additional 32-bit displacement field). The instructions are listed in table 13.

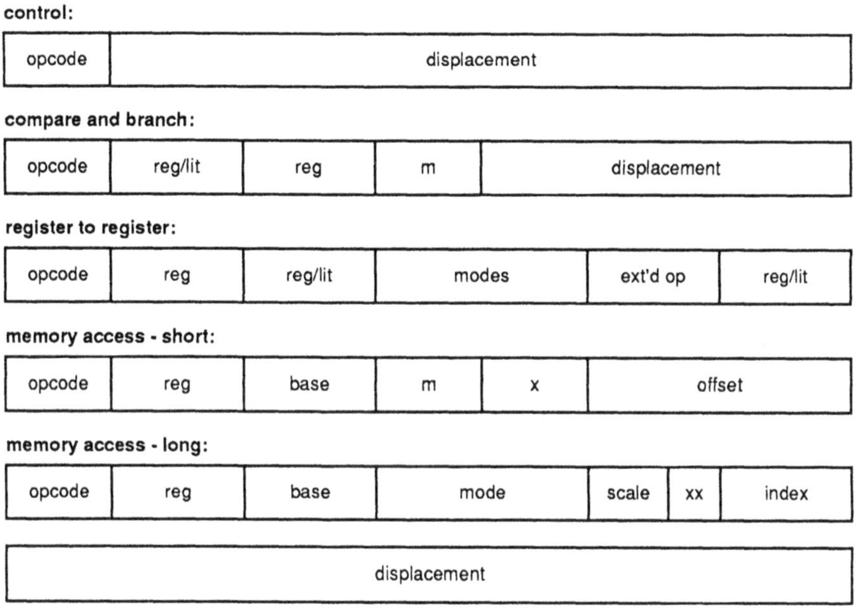

control:

opcode	displacement

compare and branch:

opcode	reg/lit	reg	m	displacement

register to register:

opcode	reg	reg/lit	modes	ext'd op	reg/lit

memory access - short:

opcode	reg	base	m	x	offset

memory access - long:

opcode	reg	base	mode	scale	xx	index

displacement

Figure 83: 80960KB instruction formats
(From Intel, 1988)

The LOAD instructions use a delayed mechanism with a register scoreboard
to provide interlocks; up to three LOADs may be pending at a time. The
following addressing modes are supported:

- 12- or 32-bit offset.
- Register indirect.
- Register plus 12- or 32-bit offset.
- Register plus (index x 1, 2, 4, 8 or 16).
- (Register x 1, 2, 4, 8 or 16) plus 32-bit displacement.
- Register plus (index x 1, 2, 4, 8 or 16) plus 32-bit displacement.

The above instruction set and addressing modes are extensive by RISC
standards.

Table 13. The Intel 80960KB instruction set

Data Movement Instructions

Load, Store, Move, Load Address.

Arithmetic Instructions

Add, Subtract, Multiply, Divide, Remainder, Modulo, Shift,
Extended Multiply and Divide.

Logical Instructions

And, Not And, And Not, Nand, Or, Not Or, Or Not, Nor,
Exclusive Or, Exclusive Nor, Not, Rotate.

Bit and Bit Field Instructions

Set bit, Clear Bit, Not Bit, Check Bit, Alter Bit,
Scan for Bit, Scan Over Bit, Extract, Modify.

Comparison Instructions

Compare, Conditional Compare,
Compare and Increment, Compare and Decrement.

Branch Instructions

Unconditional Branch, Conditional Branch, Compare and Branch.

(continued)

Table 13. (continued)

Call/Return Instructions

Call, Call Extended, Call System, Return, Branch and Link.

Decimal Instructions

Move, Add with Carry, Subtract with Carry.

Floating-Point Instructions

Move Real, Add, Subtract, Multiply, Divide, Remainder, Scale, Round, Square Root, Sine, Cosine, Tangent, ArcTangent, Log, Log Binary, Log Natural, Exponent, Classify, Copy Real Extended, Compare.

Conversion Instructions

Convert Real to Integer, Convert Integer to Real.

Synchronous Instructions

Synchronous Load, Synchronous Move.

Miscellaneous Instructions

Atomic Add, Atomic Modify, Flush Local Registers, Modify Arithmetic Controls, Scan Byte for Equal, Test Condition Code.

(continued)

Table 13. (continued)

Debug Instructions

Modify Trace Controls, Mark, Force Mark.

Fault Instructions

Condition Fault, Synchronize Faults.

(Source: Intel, 1988)

3.9.2 Organization

The organization of the 80960 chip is shown in block diagram form in figure 84. Note in particular that there are two instruction decoding mechanisms. Frequently used, simple instructions are directly decoded to execute in a single cycle. More complex (and presumably less frequent) instructions are controlled by a micro-instruction ROM and a micro-sequencer. This is a blend of RISC technology for performance and CISC technology for functionality. The micro-instruction words resemble the external instruction formats, so the micro-instruction ROM is approximately an on-chip subroutine library.

There is a 512 byte on-chip instruction cache, which is direct mapped and uses efficient 16 byte block loads when a miss ocurs. An instruction prefetch unit tries to anticipate the CPU's requirements by fetching ahead so that the CPU does not usually have to wait for instructions. The instruction decoder calculates branch targets early and communicates the result to the instruction prefetcher, and as a result some branches can effectively take zero time.

The memory bus interface supports burst transfers up to 53.3 Mbyte/s, and can buffer up to three memory requests in either direction. It can transfer up to 16 bytes in a single burst. The memory bus uses multiplexed data and addresses.

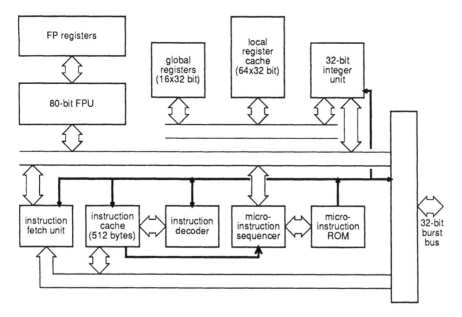

Figure 84: 960 series organization
(From Intel, 1988)

There is an on-chip floating-point unit, which supports IEEE 754 single precision, double precision, and an 80-bit extended precision. It can operate concurrently with the integer execution unit. A later version of the chip (the 80960KA) will be produced without the FPU for saving cost in applications which do not require high performance floating-point.

An unusual feature of the chip is that it has full self-test on chip. All the major functional blocks are tested at reset by a program in the micro-instruction ROM (taking 47,000 clock cycles). If a failure is located, the failure pin is asserted and execution ceases.

3.9.3 Conclusions

The 80960 architecture includes many RISC ideas, but it is not a simple chip. It has a load/store architecture, single cycle execution of many instructions, and a large register file with hardware support for procedure call and return. Against that, it has many complex instructions which make recourse to microcode, and the independent concurrent operation of several function units (the bus controller, instruction prefetcher, instruction decoder, integer execution unit and the floating-point unit) is more reminiscent of a complex mainframe architecture than a simple, pipelined RISC.

Note that this last point should not be taken as a criticism; the user should only consider price, performance, and utility when judging the suitability of a CPU. He should not care whether it is CISC, RISC, or something completely different!

As a CPU for embedded applications the 80960 is powerful and flexible. The lack of memory management will keep it out of the general-purpose CPU market for the time being.

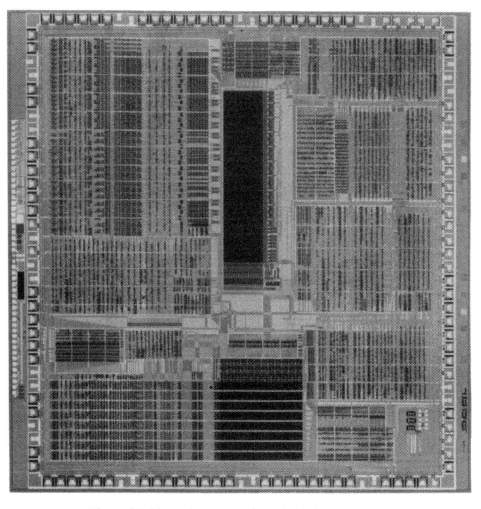

Figure 85: Photomicrograph of the C100 CLIPPER CPU
(Copyright ©1987, 1988 Intergraph Corporation.
All rights reserved.)

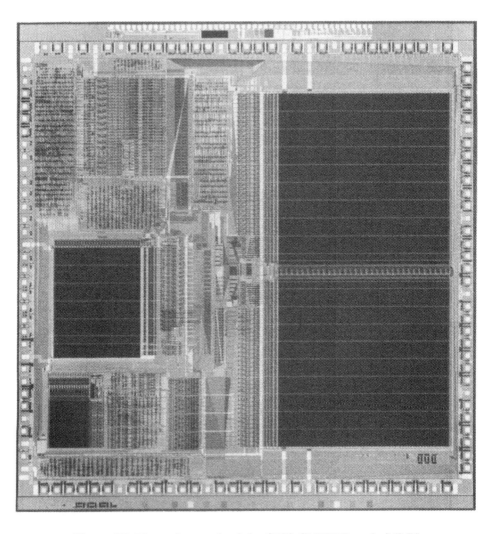

Figure 86: Photomicrograph of the C100 CLIPPER cache/MMU
(Copyright ©1987, 1988 Intergraph Corporation.
All rights reserved.)

3.10 THE INTERGRAPH CLIPPER

The CLIPPER microprocessor is based on a modified Harvard architecture which is integrated onto three chips. One chip contains the CPU with an on-chip floating-point unit, and the other two chips are identical cache and memory management units (CAMMUs), one connected to the CPU instruction bus and the other to the data bus.

The CLIPPER architecture was influenced by the RISC research at IBM, Berkeley and Stanford, but it is not itself a pure RISC. It is described by the manufacturers (Fairchild, 1986) as representing a balance between the RISC and CISC approaches.

All three chips are fabricated on a 2 micron double-metal single-polysilicon process. The CPU uses 132,000 transistors, and each CAMMU uses 357,000 transistors. With a 33MHz main clock the chip set delivers 5 MIPS sustained performance (33 MIPS peak), and higher speed versions are available.

3.10.1 Architecture

The following data types are supported:

*   Signed and unsigned bytes, halfwords, words and longwords (doublewords)

*   IEEE standard single-precision (32-bit) and double-precision (64-bit) floating-point numbers.

The CPU registers are shown in figure 87. There are 16 general-purpose registers for the user, and another 16 for the supervisor. Eight 64-bit registers are available for floating-point operands, and the two status words and the program counter complete the user visible state.

CLIPPER instructions use one to four 16-bit halfwords (figure 88). They specify two register addresses for general operations, which reduces the number of bits in the instruction compared with a more usual RISC three address format, but somewhat restricts operand re-use. The register fields are in variable positions in the instructions, though always in one of the low nibbles in a halfword.

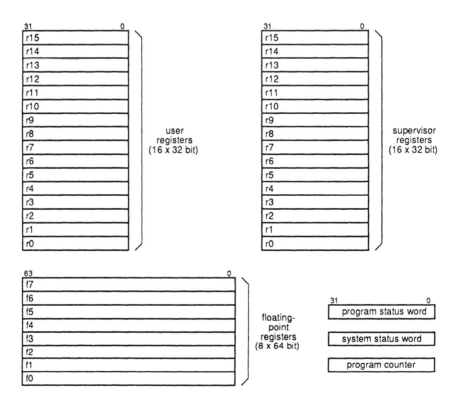

Figure 87: CLIPPER registers
(From Fairchild, 1986)

The following addressing modes are available:

- Register indirect.
- Base register plus 16- or 32-bit immediate offset.
- 16- or 32-bit immediate absolute address.
- PC plus 16- or 32-bit immediate offset.
- Base register plus index register.
- PC plus index register.

The instruction set includes support for the operations listed in table 14.

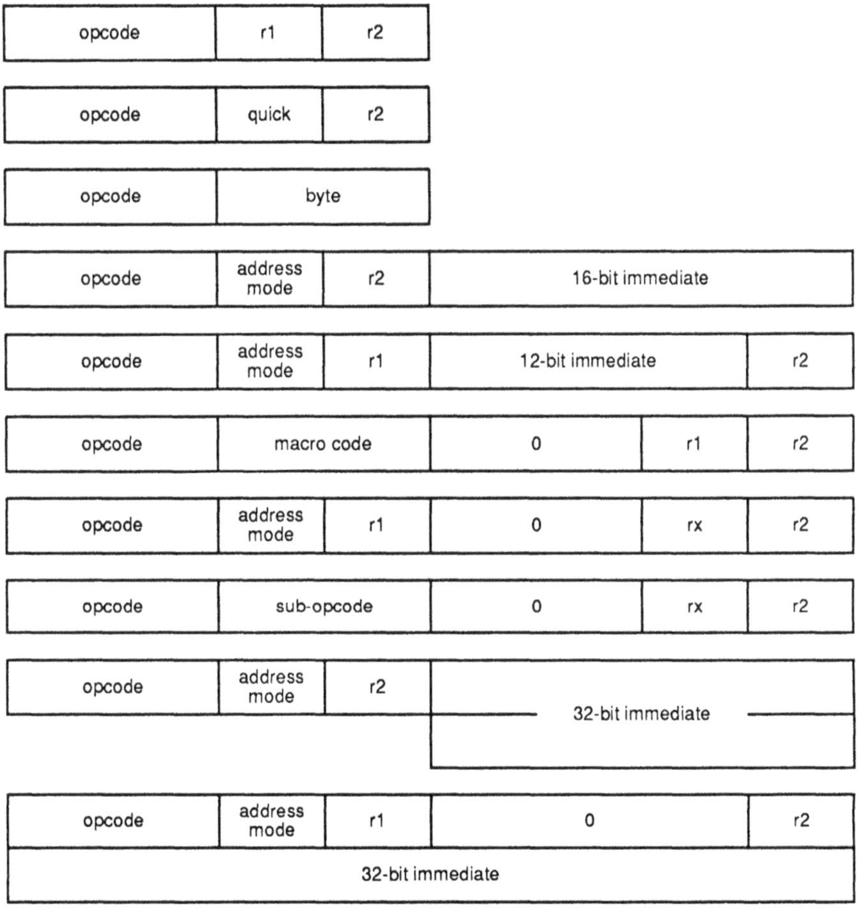

opcode	r1	r2

opcode	quick	r2

opcode	byte

opcode	address mode	r2	16-bit immediate

opcode	address mode	r1	12-bit immediate	r2

opcode	macro code	0	r1	r2

opcode	address mode	r1	0	rx	r2

opcode	sub-opcode	0	rx	r2

opcode	address mode	r2	32-bit immediate

opcode	address mode	r1	0	r2
32-bit immediate				

Figure 88: CLIPPER instruction formats
(From Fairchild, 1986)

Table 14. The CLIPPER instruction set

Data Transfer Operations

load/store   *(bytes, halfwords, words, floating-point)*
move   *(words, longwords, floating-point; between registers)*
load address

Integer Arithmetic and Logical Operations
(second operand is a register or a variable length immediate value)

add, subtract   *(with or without carry)*
multiply   *(signed, unsigned, extended)*
negate, compare
divide, modulus   *(signed and unsigned words)*
and, or, xor
not

Shift Operations

rotates and arithmetic and logical shifts of words and longwords

Floating-Point Operations
(single- or double-precision operands)

add, subtract
multiply, divide
compare, negate
scale   *(multiply by a power of two)*

(continued)

Table 14. (continued)

Conversion Operations

between integers and single- or double-precision floating-point numbers

String Operations

compare character strings
move character strings
initialize character strings

Stack Operations

push
pop
save/restore multiple registers

Control Operations

branch
call
call supervisor
call macro
return
atomic test and set
trap on floating unordered
wait for interrupt
no-op

(Source: Fairchild, 1986)

3.10.2 Organization

The organization of the CLIPPER CPU is shown in figure 89. Of particular note is the on-chip macro instruction ROM, which contains sequences of standard CLIPPER instructions which are called when a macro instruction is executed. It operates like a fixed subroutine library with a particularly efficient call mechanism.

The register bank contains sixteen user registers and sixteen supervisor registers as specified in the programming model; it also has an additional twelve registers specifically for use by the macro instructions. Likewise the floating-point register set contains the eight user/supervisor registers and four registers for macro instruction temporary use.

The integer ALU is used both for address computations and data manipulations. The shifter is a serial double bit shifter. The floating-point unit supports 64-bit IEEE standard numbers, and can operate concurrently with the integer unit.

The CLIPPER has a five stage pipeline with one instruction following one stage behind another:

(1) *Fetch.* The instruction is fetched from the cache or the macro ROM. The instruction cache has a duplicate of the PC, and may fetch ahead of demand.

(2) *Decode.* The instruction is decoded and checked to see if the required resources are free.

(3) *Operand fetch.* Integer operands are fetched from the register bank, immediate fields are extracted from the instruction.

(4) *Execute.* The ALU or the shifter performs the required operation.

(5) *Result write.* The result is written back to the register bank or memory.

Forwarding logic is used to allow an instruction to use the result of the previous operation. Only one instruction may use the floating-point unit at any time.

The CAMMUs

The cache and memory management units each have a 128 entry TLB, and perform double-level page table look up in hardware. Both the 4Kbyte cache and the TLB are dual-set associative (with least recently used set selection), so that they can both be accessed in parallel. The cache has a quadword line

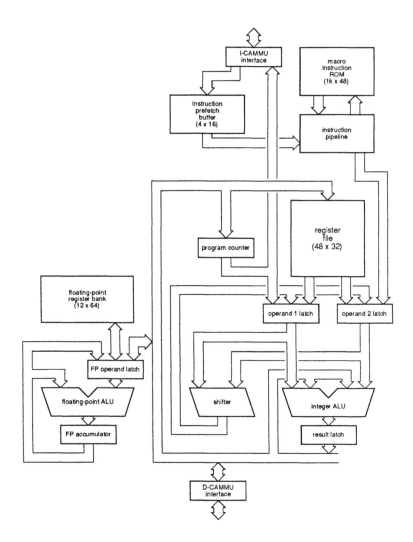

Figure 89: CLIPPER CPU organization
(From Fairchild, 1986)

length, and uses quadword block transfers for reloading. The instruction cache prefetches a quadword ahead of CPU demand.

The cache architecture supports both copy back and write through strategies, on a page by page basis (specified in the page table entry). The cache also watches the bus, and automatically updates any entries which are copies of written memory locations. It also intercepts read requests for memory locations where it is operating in copy-back mode and supplies the data itself, since the main memory may be out of date.

The CLIPPER Module

The CLIPPER module is a circuit board containing a CLIPPER CPU chip, two CAMMUs and a clock generator. It is intended to be plugged into a mother board which has the main memory and I/O functions. The system diagram is shown in figure 90.

Note that the I/O subsystem is shown containing a separate I/O processor which handles I/O transfers, thereby freeing the CLIPPER processor from the overhead of real time I/O activity. This additional processor is not a requirement, but its inclusion follows mainframe and supercomputer practice of having intelligent I/O processors rather than continually interrupting the main processor.

3.10.3 Conclusions

The designers of the CLIPPER adopted a load/store architecture, a large register bank, pipelined execution and hardwired instruction decoding from the RISC approach. Variable length instructions and multiple instruction formats, several addressing modes and complex instructions to support stacks and character strings are not characteristics of a RISC.

The system organization of the CLIPPER with its twin CAMMUs is very advanced. The memory management is conventional, and the caches large enough to be reasonably effective. The integration of the complete modified Harvard architecture onto three chips (plus a clock chip) was a considerable achievement when CLIPPER was first announced, and this three chip configuration has now been adopted by other manufacturers.

CLIPPER is only partly based on RISC principles, but it is a high-speed processor with a general-purpose system architecture well suited to UNIX workstation applications.

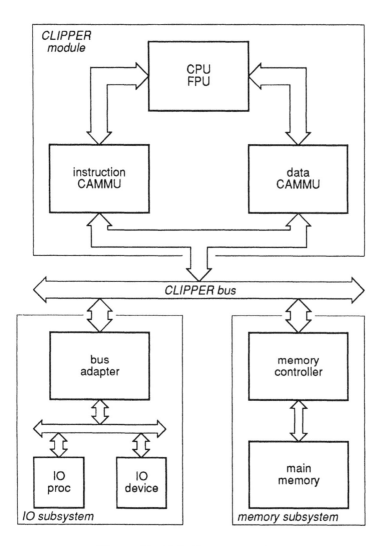

Figure 90: CLIPPER system diagram
(From Fairchild, 1986)

Figure 91: Photomicrograph of the IMS T800 floating-point transputer
(Copyright ©INMOS Limited, 1986.
Reproduced with permission of INMOS Limited.)

3.11 THE INMOS T800 TRANSPUTER

The Inmos transputer family comprises several 16- and 32-bit microprocessor configurations. They are characterized by on-chip systems for interprocessor communication aimed at enabling the construction of large-scale multiprocessors for fine-grain parallel computation. They also have substantial quantities of on-chip memory.

The transputer is designed to support the process model of computation, where a process is a black box with inputs from and outputs to other processes (which may or may not be on other processors). The Occam language has been developed to allow the concurrency implicit in the process model to be expressed efficiently, and the transputer is a VLSI Occam engine.

The transputer is probably the only microprocessor described in this book which owes nothing to the academic research described in chapter 2. However, some of the design principles are along the same lines as the RISC research. It is interesting to note both the similarities and the differences between the transputer architecture and the features of RISC designs.

The common factor between the transputer and RISC is the drive for a simple high performance processor. A RISC is simple because complexity is observed to bring diminishing returns; the transputer is simple because the CPU can only be allowed to occupy a small part of the chip area. Thereafter, most things turned out differently. The transputer uses complex microcode, has very few registers, and uses a dense byte encoded instruction set. All these characteristics run counter to the RISC philosophy.

The T800 uses 300,000 transistors on a 1.5 micron CMOS process which has one metal layer plus a low-resistivity silicide interconnect layer. The 20MHz part delivers around 10 MIPS and 1 to 2 MFLOPS.

3.11.1 Architecture

The T800 is a 32 bit microprocessor with 4Kbytes of on-chip static memory, four serial communication links, and an on-chip floating-point unit (Homewood et al, 1987). The instruction set is designed to support the *Occam* high level language efficiently, though compilers for other languages are also available.

The transputer architecture (Inmos, 1985) supports process switching at a very low level. A microcoded process scheduler on the transputer timeslices between processes automatically every 800 microseconds. Processes may operate at one of two priority levels, where high priority tasks are expected to be active for a short time, otherwise they can shut out lower priority processes.

A process which is waiting for an external event (either from the event input itself, or the arrival of data through a communication link) is suspended. If the suspended process has high priority, it will be restarted immediately when the event occurs. This is the transputer interrupt mechanism.

The programmer visible state in the CPU is contained in six 32-bit registers and 3 64-bit floating-point registers (figure 92). Three of the registers (A, B, C) form an integer evaluation stack, which shunts up and down automatically as operands are loaded or stored. The compiler is responsible for handling stack overflows and underflows. The remaining 32-bit registers are the program counter, a workspace (frame) pointer, and an operand register which is used to allow the construction of long immediate values. The small amount of state makes context switching very efficient. Interrupt latency is less than four microseconds, and procedure call and task switch are very fast too.

Floating-point operands are held on a separate 64-bit evaluation stack, which operates in a similar manner to the integer evaluation stack. There is one floating-point stack for each of the two priority levels, so that a high priority task can be switched in without saving the floating-point registers, so the addition of the FPU does not impact interrupt latency.

Transputers have just one instruction format - a single byte which is divided into one nibble of opcode and one nibble of data (figure 93).

The data is copied into the low nibble of the operand register, and the opcode actions are listed in table 15.

All instructions except the prefixing operations finish by clearing the operand register. Therefore the 16 words above the workspace pointer can be accessed in a single instruction, whereas those within 255 locations either way can be accessed in two instructions (a prefix operation and a local load or store).

The indirect operations all use implicit register addressing. The 16 most common ones use a single instruction, and all the others need a single prefix operation in addition. Multiple prefixing operations allow the instruction set to be extended indefinitely, but at present most of the possible single prefix instructions are unused.

Note how the implicit addressing of the stack model allows most of the frequently used instructions to be encoded in a single byte. A typical RISC instruction has three 5-bit register fields just to specify the sources and

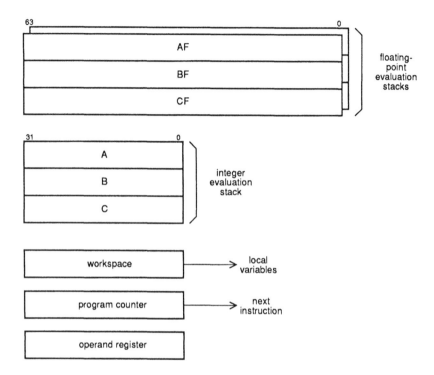

Figure 92: T800 transputer CPU registers
(From Inmos, undated)

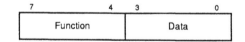

Figure 93: The transputer instruction format
(From Inmos, undated)

destination, all of which are implicit on the transputer.

The prefixing operation is also an efficient way of encoding less frequently used operations and addresses. The most common immediate fields are 0,1,-1, etc., all of which can be encoded in the single instruction. The less

Table 15. The T800 instruction set

Direct Operations

ldc	load constant	*(move operand to A, push ABC stack)*
adc	add constant	*(add operand to A)*
ldl	load local	*(load from [workspace + operand] to A,* *push ABC stack)*
stl	store local	*(store to [workspace + operand] from A,* *pop ABC stack)*
ldlp	load local pointer	*(workspace + operand offset -> A)*
ldnl	load non-local	*(load from [A + operand] to A,* *push ABC stack)*
stnl	store non-local	*(store to [A + operand] from A,* *pop ABC stack)*
ldnlp	load non-local pointer	*(A + operand offset -> A)*
ajw	adjust workspace	*(move workspace pointer by operand)*
eqc	equals constant	*(compare A with operand,* *set A=1 if equal, else 0)*
cj	conditional jump	*(jump PC relative if A=0)*
j	jump	*(jump PC relative, with task rescheduling)*
call	procedure call	*(PC relative, save ABC & PC to workspace)*

Prefixing Operations

pfix	prefix	*(shift operand left four bits)*
nfix	negative prefix	*(complement operand and shift left 4 bits)*

Indirect Operation

opr	operate	*(use operand register as extended opcode)*

(continued)

Table 15. (continued)

Single Byte Extended Operations

rev	*reverse*	*(swap A and B)*
lb	*load byte*	*(load byte addressed by A into A)*
bsub	*byte subscript*	*(address B bytes from A)*
wsub	*word subscript*	*(address B words from A)*
diff	*difference*	*(subtract without overflow and carry)*
sub	*subtract*	*(subtract with overflow and carry)*
add	*add*	*(add with overflow and carry)*
prod	*product*	*(multiply without overflow or carry)*
gt	*greater than*	*(if B>A set A=1, else set A=0)*
gcall	*general call*	*(swap A and PC)*
startp	*start process*	*(at PC+B with workspace at A)*
endp	*end process*	*(successor process workspace at A)*
outbyte	*output byte*	*(send byte in A down channel B)*
outword	*output word*	*(send word in A down channel B)*
in	*input message*	*(A bytes from channel B to memory at C)*
out	*output message*	*(A bytes to channel B from memory at C)*

Double Byte Extended Operations

Here there is a large set of instructions, including:

double-word shifts
two dimensional block copies
floating-point operations
CRC calculations
timer control
reverse bits in word
scheduler queue management, etc.

(Source: Inmos, 1988)

commonly used numbers require a single prefix operation (which allows -256 to +255 to be generated, using positive or negative prefixing) and a third prefix operation allows 12-bit constants to be generated. A full 32-bit constant requires 7 prefixing operations, and is probably better created as a constant to be loaded from memory.

On-chip Memory

All transputers have substantial on-chip static memory. The T800 has 4Kbytes. This memory is not organized as a cache, it is a directly addressed part of the processor's memory space. Off-chip memory is also supported by general-purpose transputers, but has poorer access time, so careful allocation of the on-chip memory is important. The on-chip memory may be used in a manner similar to the register bank of a conventional RISC (though there is much more memory on a transputer than in a RISC register bank), or as a software maintained cache of the most frequently used data values or code sequences.

Communications

Inter-process communications are an essential part of the transputer computing model, and the T800 has four *Links* which support synchronous communications between processes on different transputers. Each Link is capable of sending and receiving messages autonomously to and from another transputer, without CPU intervention except to initiate the transfer. The messages are copied between the memories of the two devices.

3.11.2 Organization

The internal organization of the T800 is shown in figure 94.

The transputer chip embodies several interesting bits of technology:

(1)   All transputers accept a 5MHz system clock, and generate higher clock rates on chip. Therefore a single clock can supply a system of mixed performance transputers, or a particular transputer can be upgraded to a higher speed part without changing the clock.

(2)   All the communications Links operate at 10Mbit/s (more recently 20Mbit/s), so transputers of different speeds can communicate easily. The Links operate asynchronously, with all the synchronization being performed on the transputer itself, so communicating transputers need not operate off the same clock (though they must all use the same system clock frequency). Similarly, if several transputers do use the

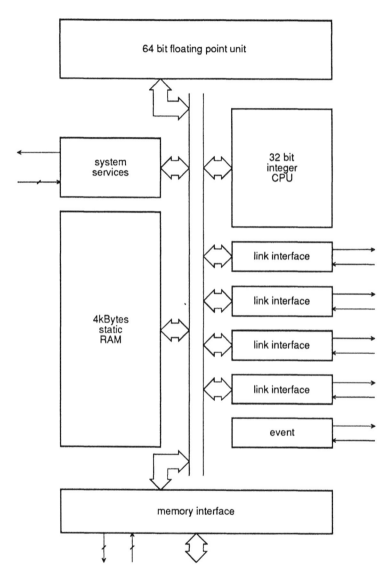

Figure 94: T800 transputer block diagram
(Adapted from Inmos, 1985)

same system clock, clock skew between them will not cause a problem.

(3)   The memory interface contains logic for controlling dynamic RAM directly, with programmable timing characteristics for devices of different speeds.

The floating-point unit (figure 95) contains the floating-point evaluation stack and the execution logic. The design of this unit was based on careful consideration of the effect of various trade-offs on performance. For instance, it would be possible to put faster floating-point hardware on a separate coprocessor chip, but this would have higher cost, and with the overhead of inter-chip communications would not give better system performance.

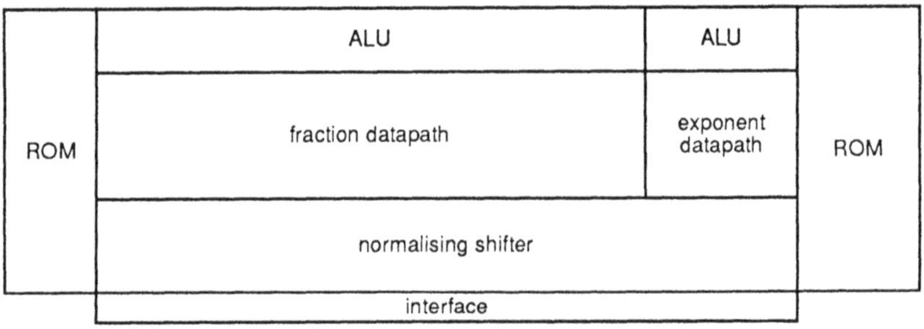

Figure 95: T800 floating-point unit

The T800 FPU performs multiplies three bits at a time, and divides two bits at a time. The normalizing shifter can normalize a result in two cycles. Transcendental functions are performed in software using polynomial approximations, but square roots are implemented in hardware in order to meet the stringent requirements of the IEEE specification.

The correct implementation of the FPU design was ensured by the use of a formal specification language and the mathematical derivation of other representations of the logic (see Homewood et al, 1987). The formal approach to the design of complex hardware systems has gained considerable favour at Inmos.

3.11.3 Conclusions

The transputer was born out of a different world view from the other CPUs described in this book. Though there is a common aim of producing a simple, fast CPU, the transputer embodies a solution to this problem which is very different from that adopted by the designers of the other processors.

The process model of computation, which is the basis of Occam and the transputer architecture, is quite different from the model upon which the design of most of the world's software is based. The transputer is not optimized for use as the CPU in a UNIX workstation running large Fortran programs. It has no support for memory management, and no protection for one task against another. (If you want a protected process on a transputer, you should run it on a separate transputer.)

However, the transputer is well matched to a number of other application areas. At the high end, if performance is all-important, many applications can be rewritten to allow a large network of transputers to operate with high levels of concurrency. The transputer is a unique VLSI component for the construction of fine-grain parallel supercomputers.

At the other extreme, the transputer is finding many applications as a high performance processor for embedded controllers. It is remarkably easy to bring up a transputer based system, using a separate Link adaptor on a PC card to communicate with the transputer for program development. No bootstrap ROM is required, as the transputer can be configured to accept the first Link message as a bootstrap program. The dynamic RAM interface on the chip allows a very compact computing node to be built, and the small number of registers makes real-time performance good. If you should run out of processing power, adding another transputer is also quite straightforward!

Other versions of the transputer are available or in development. The T424 is a T800 without the floating-point unit. The T414 is similar to the T424, but has only 2Kbytes of on-chip RAM. 16-bit transputers are available, and some have the general-purpose memory interface removed and replaced by a dedicated peripheral controller and interface. All can communicate with each other via the Links.

The use of formal design methodologies to ensure the correct operation of the T800 floating-point unit represents a significant step forward. The multi-million transistor chips of the near future cannot be designed correctly without substantial advances in design methodologies. Exhaustive testing of designs is already impractical. Mathematical approaches, such as that used to prove that the T800 FPU implementation corresponds to the specification, offer substantial leverage in this area.

References

AMD, (1987). Am29000 User's Manual, Advanced Micro Devices.

Ditzel, D. R., McLellan, H. R. and Berenbaum, A. D., (1987). "The Hardware Architecture of the CRISP Microprocessor," Proceedings of the 14th Annual Symposium on Computer Architecture, pp. 309-319.

Ditzel, D. R. and McLellan, H. R., (1982). "Register Allocation for Free: The C Machine Stack Cache," Proceedings of the Symposium on Architectural Support for Programming Languages and Operating Systems, pp. 48-54.

Ditzel, D. R. and McLellan, H. R., (1987). "Branch Folding in the CRISP Microprocessor: Reducing Branch Delay to Zero," Proceedings of the 14th Annual Symposium on Computer Architecture, pp. 2-9.

Dobbs, C., Reed, P. and Ng, T., (1988). Supercomputing on a Chip - Genesis of the 88000, VLSI System Design, May, pp. 24-33.

Fairchild, (1986). Introduction to the CLIPPER Architecture, Fairchild Camera and Instrument Company, Palo Alto, California. (CLIPPER manuals are now available from Intergraph Advanced Processor Division, Embarcadero Place, Building D, 2400 Geng Road, Palo Alto, CA 94303.)

Forsyth, M., Jaffe, W. S., Tanksalvala, D., Wheeler, J. and Yetter, J., (1987). A 32-bit VLSI CPU with 15-MIPS Peak Performance, IEEE Journal of Solid-State Circuits, sc-22, no. 5, pp. 768-775.

Fujitsu, (1987). MB86900 High Performance 32-bit RISC SPARC data sheet, Fujitsu Microelectronics Inc.

Hewlett-Packard, (1986). Precision Architecture and Instruction Reference Manual, Hewlett-Packard Company.

Homewood, M., May, D., Shepherd, D. and Shepherd, R., (1987). The IMS T800 Transputer, IEEE Micro, 7, no. 5, pp. 10-26.

IBM, (1986). IBM RT Personal Computer Technology, IBM Corporation, Austin, Texas.

Inmos, (1985). Transputer Reference Manual, Inmos Ltd, Bristol, England. (New edition available from Prentice Hall International (UK) Ltd, Hemel Hempstead, Hertfordshire, England.)

Inmos, (1988). The Transputer Instruction Set: a compiler writers guide, Prentice Hall International (UK) Ltd, Hemel Hempstead, Hertfordshire, England.

Inmos, (undated). The Transputer Implementation of Occam, Technical Note 22, Inmos Ltd, Bristol, England.

Intel, (1988). 80960KB Embedded 32-bit Microprocessor with Integrated Floating-Point Unit, Intel Corporation.

Kane, G., (1987). MIPS R2000 RISC Architecture, Prentice Hall, Englewood Cliffs, N.J. 07632.

Mahon, M. J., Lee, R. B-L., Miller, T., Huck, J. C. and Bryg, W. R., (1986). Hewlett-Packard Precision Architecture: The Processor, Hewlett-Packard Journal, 37, no. 8, pp. 4-22.

Motorola, (1988). MC88100 & MC88200 Technical Summaries, Motorola Inc., Austin, Texas.

Radin, G., (1983). The 801 Minicomputer, IBM Journal of Research and Development, 27, no. 3, pp. 237-246.

Sun, (1987). The SPARC Architecture Manual, Sun Microsystems Inc., Mountain View, California.

Sun, (1988). The SPARC Reference MMU, Sun Microsystems Inc., Mountain View, California.

VSLI Technology, Inc., (1987). VL86C010 RISC family data manual, Application Specific Logic Products Division, 8375 South River Parkway, Tempe, AZ 85284.

4

The Implementation of the ARM

The Acorn RISC Machine is significantly simpler than most other VLSI RISC implementations. An overview of the architecture of the ARM processor was presented in the last chapter. In chapter 5 there is a description of the ARM3, which combines the ARM2 CPU cell with a 4Kbyte on-chip mixed instruction and data cache.

In this chapter we will look into the design of such a processor. The ARM is a good introduction to the design issues because of its relative simplicity, and a detailed description of the internal operation of the CPU is presented here. Design methodologies will also be discussed; the methodologies evolved during the project, and had to be considered with as much care as the VLSI components themselves.

The design of the ARM started with the specification of the instruction set. From this a datapath organization was determined, and then a simulation model built. The simulation model was subjected to extensive testing, and then itself used to generate test patterns for the circuit-level implementation. We shall look at all these stages of development in this chapter.

4.1 INSTRUCTION SET AND DATAPATH DEFINITION

The ARM instruction set was the first aspect of the processor to be specified. It was designed by Roger Wilson during the latter half of 1983, and he drew mainly on his own experience of writing interpreters, together with a good insight into the needs of compiler writers, to determine the level of functionality. Published studies of instruction distributions in large compiled programs were used for reference, but no specific studies were undertaken for this project. The instructions were slightly modified in the course of developing the hardware design, but no major changes were involved.

The instruction set designer is faced with a very large set of options, indeed the possible choices are unbounded. Normally a number of simplifying decisions are made early in the design, with or without justification. There is a clear trade-off between the semantic density of an instruction set and the simplicity of the hardware required to execute the instructions. In principle an analysis could be made of instruction set usage, and a very dense instruction set produced by frequency encoding the instructions. In practice, even CISC designers restrict their instructions to formats based on multiples of a byte (or whatever their choice of basic storage unit is). The RISC trend is to swing even more towards regular instruction formats and simple hardware.

The design of the ARM instruction set started with a decision to use a single instruction length of exactly one word (32 bits), and the load/store model was adopted. Both of these decisions were based on evidence from the early RISC research described in chapter 2, and have been widely (though not universally) followed by other RISC designs. This provided sufficient framework to allow other decisions to be made on an instruction by instruction basis, and the issues will be discussed later in this section as the instruction set is described in detail.

The ARM instruction set has been extended with each version of the processor. The basic set used on the original 3 micron part (ARM1) will be described first, then the additions which were included on the two micron part (ARM2). The additions were based on experience with the ARM1 instruction set, and (particularly in the case of the multiply instruction) the reasoning behind the change throws light on some issues which were not obvious at the time of the first design. The semaphore instruction which was added to ARM3 (the CPU with on-chip cache) is covered in the section on ARM3 in the next chapter, and will not be discussed here as it was not included in either version of the uncached CPU.

Once the basic design of the instruction set was satisfactory, the hardware design was begun. Here the chosen approach was to match the processor very closely to the characteristics of DRAM (dynamic random access memory), which is by far the most cost-effective bulk semiconductor memory available. The von Neumann model was taken as the most effective for a low cost design, so that instructions and data share a single 32-bit port to main memory. A direct consequence of this decision is that LOAD and STORE instructions must take more than one cycle, since there will always be one instruction and at least one data value to be transferred. This has the beneficial effect of allowing full base plus index (or offset) with auto-increment or decrement addressing modes to be implemented with only two register read ports. A single cycle STORE would require a Harvard architecture, and either three read ports or a simpler addressing mode.

A characteristic of DRAMs is that they support faster accesses when the address stays within the same row in the memory array. Processors typically use sequential addresses a lot of the time, but then switch unpredictably to unrelated addresses. A sequential address arrives too late for a memory controller to use it to decide whether it can safely use the faster mode. After considerable thought about how best to use the faster memory modes, a simple solution was adopted. The processor requires an address incrementer so that the ALU is free to do processing work during normal sequential instruction execution; a control signal was taken off chip to flag when the address for the next cycle is to be taken from this incrementer. A memory controller can then check the old address to detect when a page boundary is about to be crossed, and it can use this information together with the sequential indicator to decide whether the new address will be suitable for a fast memory access mode to be used. This decision can be made well before the new address becomes valid.

The datapath was designed by inspection of the instruction set, with the aim of implementing all the functions required by each instruction in the smallest possible number of cycles (usually determined by the number of memory accesses required) with the simplest bus structure that would work. Draft datapath designs were produced and photocopied to produce several blanks; the flow of data during each cycle of each instruction was then drawn onto a blank to check for clashes of resource requirement. The design was iterated until it was felt that an optimal point had been reached. The final data flow diagrams are reproduced later in this chapter, along with the descriptions of the individual instructions.

This graphical representation of data movement was practical for ARM because there are so few instruction types. It would not be satisfactory for a much more complex instruction set, as the paper area required to represent all the instruction types would become unmanageable. A more formal approach to the data flow management is required for complex processor design. An example of a formal approach is described by Tredennick (1987).

One major decision was on the pipeline depth of the processor. It was decided not to pipeline the datapath at all, to avoid the complexities of register forwarding hardware (and we did not wish to require our compilers to manage data race hazards in software). Since we were hoping to make optimal use of DRAMs at their specified performance limits, instructions would be arriving right at the end of the access cycle, so it would not be possible to decode them in time for the next cycle. We therefore introduced a decode cycle, and adopted a three stage fetch-decode-execute pipeline.

The processor clock structure was chosen on the basis of caution and user experience with 6502 8-bit processors. The ability of the manufacturers of the 6502 to increase the clock rate four-fold over the life of the product was taken as evidence in favour of simple clocking schemes, and a two-phase per CPU cycle clock was adopted. Externally generated (unbuffered) clocks were accepted in order to be able to control the CPU functions to match external memory timing accurately. The processor does not have a 'wait' signal; instead one of the clock phases is stretched when the memory system cannot respond in the normal cycle time. This, again, was exactly how the 6502 had been controlled in earlier Acorn products.

Once these high level decisions had been made, detailed design work commenced. A simple model of the CPU was built in BASIC (for the first CPU; subsequent CPUs were based on models written in Modula-2). Each functional block on the chip was described by two procedures, one for each clock phase. We shall return to the this aspect of the design methodology after looking at the instruction set and the results of the data flow studies.

In this section we look at each instruction in detail, at the datapath activity which is required to implement each instruction, and at the expected state of the CPU pins in each cycle.

The datapath activity is shown graphically on block diagrams, with shading to indicate active buses. Each active datapath block is marked with the function that it is required to perform in the corresponding cycle.

In the accompanying tables of CPU pin activity $\overline{\text{MREQ}}$ and **SEQ** (the active-low memory request and active-high sequential cycle request signals respectively, which are pipelined up to one cycle ahead of the cycle to which they apply) are shown in the cycle in which they appear, so they may be used to predict the address of the next cycle. The address, $\overline{B}/W$, $\overline{R}/W$, and $\overline{OPC}$ (the not byte/word, not read/write and active low op-code fetch indicators, which appear up to half a cycle ahead) are shown in the cycle to which they apply. This corresponds to the way the memory controller support chip uses these signals.

We begin by examining the instruction set implemented on the original 3 micron ARM; later we shall look at the extensions added to the 2 micron version of the CPU:

4.1.1 The Condition Field

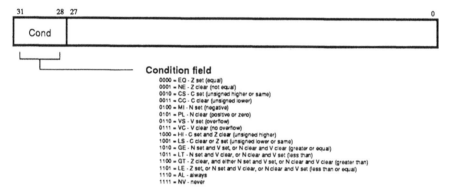

Condition field
0000 = EQ - Z set (equal)
0001 = NE - Z clear (not equal)
0010 = CS - C set (unsigned higher or same)
0011 = CC - C clear (unsigned lower)
0100 = MI - N set (negative)
0101 = PL - N clear (positive or zero)
0110 = VS - V set (overflow)
0111 = VC - V clear (no overflow)
1000 = HI - C set and Z clear (unsigned higher)
1001 = LS - C clear or Z set (unsigned lower or same)
1010 = GE - N set and V set, or N clear and V clear (greater or equal)
1011 = LT - N set and V clear, or N clear and V set (less than)
1100 = GT - Z clear, and either N set and V set, or N clear and V clear (greater than)
1101 = LE - Z set, or N set and V clear, or N clear and V set (less than or equal)
1110 = AL - always
1111 = NV - never

All ARM instructions are conditionally executed, which means that their execution may or may not take place depending on the values of the N, Z, C and V flags in the PSR at the end of the preceding instruction.

If the ALways condition is specified, the instruction will be executed irrespective of the flags, and likewise the NeVer condition will cause it not to be executed (it will be a no-op, ie take one cycle and have no effect on the processor state).

The other condition codes have meanings as detailed above, for instance code 0000 (EQual) causes the instruction to be executed only if the Z flag is set. This would correspond to the case where a compare (CMP) instruction had found the two operands to be equal. If the two operands were different, the compare instruction would have cleared the Z flag, and the instruction will not be executed.

Having all instructions execute conditionally is a feature peculiar to the ARM. It is justified partly because it offsets the relative inefficiency of the branch instruction, and partly on the basis that the condition evaluation hardware has to be there anyway for the branch instructions, so why not use it for all the instructions rather than leaving it idle most of the time? It replaces short forward branches; IF..THEN..ELSE statements can be compiled without any branches at all. This saves code space and considerable execution time, since an unexecuted instruction preserves the pipeline whereas a branch destroys it.

In typical programs a high proportion of ARM instructions use the ALways condition. Nontheless, if each conditional instruction is actually saving a branch, this approach is contributing noticeably to both code density and performance.

Condition evaluation is performed alongside operand fetch during the first phase of the first cycle of execution. If the condition fails, all write activity is suppressed, and further cycles of the failed instruction are omitted. The condition evaluation therefore has no impact on processor cycle time, but a failed instruction still takes one cycle even though it does nothing. This cycle would be necessary anyway, however, because another instruction must be fetched to fill the pipeline slot vacated by the failed instruction.

4.1.2 Branch and Branch with Link Instructions

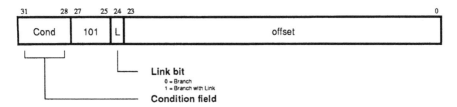

All branches take a 24-bit offset. This is shifted left two bits and added to the PC, with any overflow being ignored. The branch can therefore reach any word-aligned address within the 64 Mbyte address space. The branch offset must take account of the prefetch operation, which causes the PC to be 2 words ahead of the current instruction. The *Branch with Link* instruction copies the old PC and PSR into R14 of the current bank, for subsequent use as a subroutine return address. The PC value written into the link register (R14) is adjusted to allow for the prefetch, and contains the address of the instruction following the branch and link instruction.

The link mechanism is simpler than a return address stack in external memory, and more efficient. A leaf procedure can keep the return address in R14, thereby avoiding sending it to memory at all. Other procedures can save it to memory along with work registers using a single STM {R0,..,Rn,R14}, and return and restore the work registers using a single LDM {R0,..Rn,PC}. An external memory stack implemented in hardware would cause the sequentiality of the memory access to be broken twice upon procedure entry, once for saving the return address and once for saving the work registers, and twice upon exit. The link mechanism (together with STM/LDM) breaks sequentiality only once each time.

The branch instruction breaks the pipeline, and causes two prefetched instructions to be discarded. Delayed branching was not implemented because of the complexities of retaining multiple old PC values, particularly in the presence of re-entrant exception handlers. If the instruction in the delay slot causes a memory abort, restarting it can be very difficult. The solution usually adopted is to keep a single PC history and disallow re-entrance at least until the history has been saved. Such an approach makes use of special registers, and conflicts with using a general register (R14) for the return address both for subroutine and exception handling. We preferred the simplicity and generality of the link register approach, and accepted the performance penalty that results from this decision. Conditional code compensates somewhat by reducing the number of branches required (and performing better, when applicable, than even a delayed branch).

Branch Implementation

The datapath activity during a branch or branch and link is shown in figure 96, and the CPU pin states are shown in table 16. In the table 'pc' is the address of the branch instruction, 'alu' is an address calculated by ARM, '(alu)' are the contents of that address, etc.

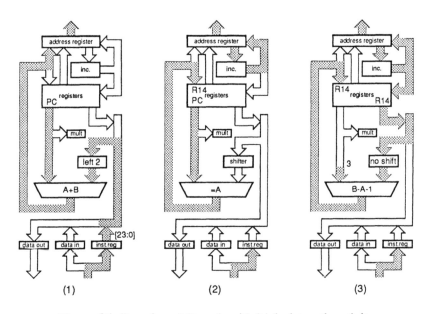

Figure 96: Branch and Branch with Link datapath activity

A branch instruction calculates the branch destination in the first cycle by shifting the bottom 24 bits of the branch instruction left 2 bits to convert them to a word offset, and then adding them to the PC (using the main ALU). At the same time, it performs a prefetch from the current PC. This prefetch is done in all cases, since by the time the decision whether or not to take the branch has been reached, it is already too late to prevent the prefetch.

Table 16. Branch and Branch with Link memory interface signals

Cycle	address	b/w	r/w	data	seq	$\overline{mreq}$	$\overline{opc}$
1	pc+8	1	0	(pc+8)	0	0	0
2	alu	1	0	(alu)	1	0	0
3	alu+4	1	0	(alu+4)	1	0	0
	alu+8						

During the second cycle a fetch is performed from the branch destination, and the old PC is copied to register 14 if the link bit is set (for subsequent use as a return address). Note that both branch and branch with link perform all the datapath activities identically with common decode structures; the link bit itself is used to differentiate the two cases by directly enabling or disabling the link register write activity in the second and third cycles.

The third cycle performs a fetch from the destination + 4, thereby refilling the instruction pipeline. The incremented destination address is copied back into R15 at the same time, to ensure that the PC has the correct value for the next instruction (where it might be used as a base address for a load or store instruction, for example). If the branch is with link, R14 is modified (4 is subtracted from it) so that a return may use MOV PC,R14. If this were not done, a return would require SUB PC,R14,#4. This would not be a problem for a leaf subroutine which can return with either of the data operations equally easily, but the modification of the link also allows a subroutine which calls other subroutines to save the link register (and some workspace registers) with a STM..{R14} instruction, and return (and restore the workspace registers) with a LDM ..{PC} instruction. The subtraction of 4 is performed by subtracting 3 with borrow, since this mechanism is required independently for LDM and STM offset calculations.

4.1.3 Data Processing Instructions

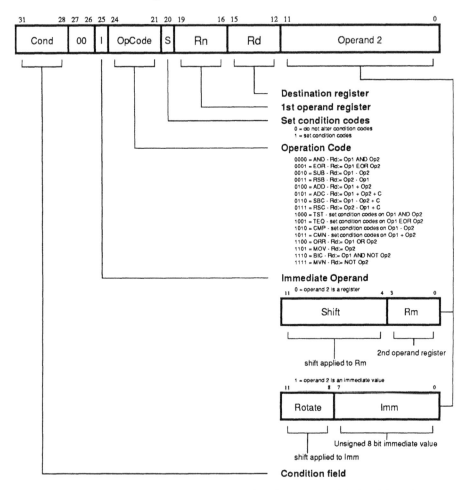

The instruction produces a result by performing a specified arithmetic or logical operation on one or two operands. The first operand is always a register (Rn). The second operand may be a shifted register (Rm) or a rotated 8-bit immediate value (Imm) according to the value of the I bit in the instruction. The condition codes in the PSR may be preserved or updated as a result of this instruction, according to the value of the S bit in the instruction. Certain operations (TST, TEQ, CMP, CMN) do not write the result to Rd. They are used only to perform tests and to set the condition codes on the result, and therefore should always have the S bit set.

This instruction class is in some ways more complex than CISC equivalents. Few CISCs perform a general shift operation and a general ALU operation in a single instruction. The high functionality of the instruction is in part a direct result of the 32-bit instruction format; when the instruction class does not demand a large immediate field (such as the 24-bit branch offset), there is an almost embarrassing number of bits left even after specifying three register numbers, and the condition and op-code fields. If the shift options were not included, it would be difficult to find a good use for all 32 bits. The shift operation turns out to be a powerful feature in practice, but its inclusion is justified only if hardware can be built to support it without significantly slowing the datapath cycle time. Fortunately, this is possible.

The shift options include left and right arithmetic and logical shifts and rotates, by 0 to 32 bits (or register specified amounts up to 255 bits). When an immediate operand is specified, it may be rotated into any even bit position. Rotating the immediate field right by six bits, for instance, allows the supervisor to set all the PSR bits (which occupy bits 0,1 and 26 to 31 of register 15).

The choice of a shifted byte immediate operand, as opposed to an unshifted 12 bit operand, was influenced by the observed distribution of immediate constants. These tend predominantly to be small positive or negative integers, with lower frequency occurrences being larger numbers which tend to be powers of two. These may all be generated by the shifted byte field. (This decision may be contrasted with the equivalent decision for the load and store offsets, where the same options were possible but the opposite decision was reached.)

The explicit control of the setting of the condition codes allows the condition codes to hold state over several instructions without requiring an unnecessary instruction to set it in the first place. Older CPUs tend either always to set the flags (and destroy the old state) every instruction, or to set it only when compares are performed. The explicit control bit in the instruction is more flexible than either approach, and is becoming widespread in newer architectures.

Data Operation Implementation

A data operation executes in a single datapath cycle except where the shift is determined by the contents of a register, or where the PC is modified by the instruction. A register is read onto the A bus, and a second register (figure 97(1)) or the immediate field (figure 97(2)) onto the B bus. The ALU combines the A bus source and the shifted B bus source according to the

Table 17. Data Operation memory interface signals

	Cycle	address	b/w	r̄/w	data	seq	m̄r̄ēq	ōpc
normal	1	pc+8	1	0	(pc+8)	1	0	0
		pc+12						
dest=pc	1	pc+8	1	0	(pc+8)	0	0	0
	2	alu	1	0	(alu)	1	0	0
	3	alu+4	1	0	(alu+4)	1	0	0
		alu+8						
shift(Rs)	1	pc+8	1	0	(pc+8)	0	1	0
	2	pc+12	1	0	–	1	0	1
		pc+12						
shift(Rs),	1	pc+8	1	0	(pc+8)	0	1	0
dest=pc	2	pc+12	1	0	–	0	0	1
	3	alu	1	0	(alu)	1	0	0
	4	alu+4	1	0	(alu+4)	1	0	0
		alu+8						

operation specified in the instruction, and the result (when required) is written to the destination register. (Compares and tests do not produce results, only the ALU status flags are written.)

An instruction prefetch occurs at the same time as the above operation, and the program counter is incremented.

When the shift length is specified by a register, an additional datapath cycle occurs before the above operation to copy the bottom 8 bits of that register

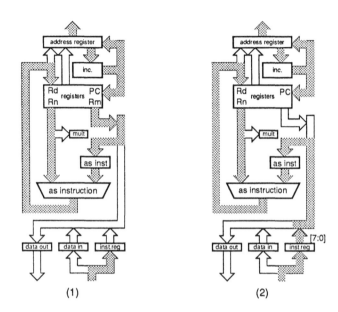

Figure 97: Data Operation datapath activity

into a holding latch in the barrel shifter. The instruction prefetch will occur during this first cycle, and the operation cycle will be internal (ie will not request memory). This internal cycle is configured to merge with the next cycle into a single memory N-cycle when MEMC is used as the memory interface.

The PC may be any (or all!) of the register operands. When read onto the A bus it appears without the PSR bits; on the B bus it appears with them. Neither will affect external bus activity. When it is the destination, however, external bus activity may be affected. If the result is written to the PC, the contents of the instruction pipeline are invalidated, and the address for the next instruction prefetch is taken from the ALU rather than the address incrementer. The instruction pipeline is refilled before any further execution takes place, and during this time exceptions are locked out.

4.1.4 Single Register Load and Store Instructions

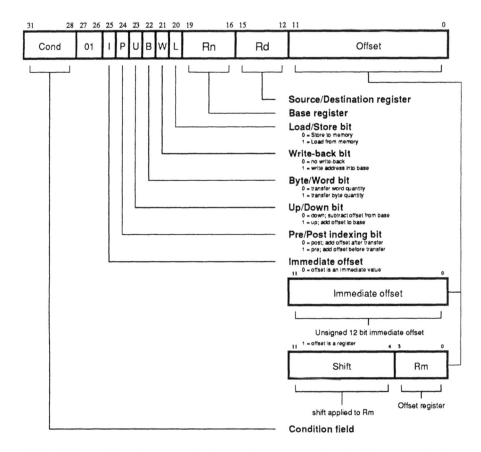

The single register data transfer instructions are used to load or store single bytes or words of data. The memory address used in the transfer is calulated by adding an offset to or subtracting an offset from a base register. The result of this calculation may be written back into the base register if 'auto-indexing' is required.

The offset from the base may be either a 12-bit unsigned binary immediate value in the instruction, or a second register (possibly shifted in some way). The offset may be added to (U=1) or subtracted from (U=0) the base register Rn. The offset modification may be performed either before (pre-indexed, P=1) or after (post-indexed, P=0) the base is used as the transfer address.

The W bit gives optional auto increment and decrement addressing modes. The modified base value may be written back into the base (W=1), or the old base value may be kept (W=0). In the case of post-indexed addressing, the write back bit is redundant, since the old base value can be retained by setting the offset to zero. Therefore post-indexed data transfers always write back the modified base.

The choice of immediate offset was different in this case from the data operation case. Here the distribution of offsets shows no tendency to include large integers which are powers of two, so the unshifted 12-bit offset is more useful than the shifted byte would be.

The flexibility of the addressing modes (compared with some other RISCs) results directly from the acceptance of the von Neumann model; an instruction and a data word must be transferred across a port which supports the transfer of one word per cycle, so the instruction has to take at least two cycles. This allows the first cycle to be used solely for address calculation, which can therefore exploit all the datapath features that are available for the data operation. The second cycle is used for the data transfer to or from memory, and any auto-indexing action required by the instruction. Since data arrives from memory very late in the cycle, it is not possible to transfer into a register in the transfer cycle, so loads use a third cycle for moving the data to the destination register. The third cycle has no time penalty in a DRAM-based system, as explained later (in the section on the memory interface).

Load Register Implementation

The first cycle of a load register instruction performs the address calculation. This may be either base plus offset (as in figure 98), or base plus index register (as shown in the Store register datapath activity figure). The index or offset may be added to or subtracted from the base, or the base may be used unmodified. The instruction decoding is the same for these options; the U and P bits in the instruction modify the ALU function directly to select the required mode.

The data is fetched from memory during the second cycle, and the base register modification is performed during this cycle (if required). The calculation is always performed; the W and P bits are used to determine whether or not the base register is written with the calculated value. The B bit in the instruction is communicated to the memory system to select a byte or word transfer. It has no effect on the processor activity until the third cycle (and could be ignored by the memory system for loads, but not for stores).

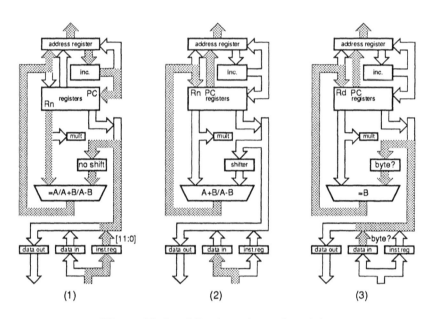

Figure 98: Load Register datapath activity

During the third cycle the data is transferred to the destination register via the shifter and the ALU, and external memory is unused. If a byte load was requested, the selected byte is extracted from the loaded word onto the B bus, and then rotated into the least significant byte position in the shifter. The extraction process fills the other byte positions with zeroes, so after rotation the byte is zero-extended to 32 bits. The ALU simply transmits the shifter output to the destination register. This third cycle may normally be merged with the following prefetch to form one memory N-cycle.

Either the base or the destination (or both) may be the PC, and the prefetch sequence will be changed if the PC is affected by the instruction. PC-relative addressing is often used to pick up address constants which can be placed near the related code, but auto-indexing is not very useful when the PC is the base.

The data fetch may abort, and in this case the base and destination modifications are prevented. The abort condition simply disables subsequent register modification until the instruction has completed.

Table 18. Load Register memory interface signals

	Cycle	address	b/w	r/w	data	seq	mreq	opc	trans
normal	1	pc+8	1	0	(pc+8)	0	0	0	
	2	alu	b/w	0	(alu)	0	1	1	t
	3	pc+12 pc+12	1	0	–	1	0	1	
dest=pc	1	pc+8	1	0	(pc+8)	0	0	0	
	2	alu	b/w	0	(alu)	0	1	1	t
	3	pc+12	1	0	–	0	0	1	
	4	(alu)	1	0	((alu))	1	0	0	
	5	(alu)+4 (alu)+8	1	0	((alu)+4)	1	0	0	
base=pc,	1	pc+8	1	0	(pc+8)	0	0	0	
write-back	2	alu	b/w	0	(alu)	0	1	1	t
dest#pc	3	pc*	1	0	–	0	0	1	
	4	pc*	1	0	(pc*)	1	0	0	
	5	pc*+4 pc*+8	1	0	(pc*+4)	1	0	0	
base=pc,	1	pc+8	1	0	(pc+8)	0	0	0	
write-back	2	alu	b/w	0	(alu)	0	1	1	t
dest=pc	3	pc*	1	0	–	0	0	1	
	4	(alu)	1	0	((alu))	1	0	0	
	5	(alu)+4 (alu)+8	1	0	((alu)+4)	1	0	0	

In table 18, 'pc*' is the PC value modified by write-back, and 't' shows the cycle where the force translation option in the instruction may be used to enable the supervisor to access the user's virtual address space.

Store Register Implementation

The first cycle of a store register is the same as the first cycle of load register. Figure 99 shows the base plus index case; the base plus immediate offset case is shown in the LDR figure. During the second cycle the base modification is performed, and at the same time the data is written to memory. There is no third cycle.

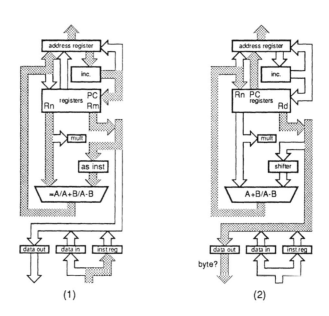

Figure 99: Store Register datapath activity

A byte store replicates the low byte of the source register four times across the data bus, and external memory control should activate only the addressed byte block of the memory.

The PC cannot be modified by this instruction unless it is the base register and write-back is enabled. Though this is allowed and supported, it is not often useful.

A data abort prevents the base write-back, as with the load register instruction. The abort handling software in a paged virtual memory system can use the data abort exception return address to locate the instruction which

Table 19. Store Register memory interface signals

	Cycle	address	b/w	r/w	data	seq	mreq	opc	trans
normal	1	pc+8	1	0	(pc+8)	0	0	0	
	2	alu pc+12	b/w	1	Rd	0	0	1	t
base=pc,	1	pc+8	1	0	(pc+8)	0	0	0	
write-back	2	alu	b/w	1	Rd	0	0	1	t
	3	pc'	1	0	(pc')	1	0	0	
	4	pc'+4 pc'+8	1	0	(pc'+4)	1	0	0	

faulted, and it must then interpret that instruction to deduce the address which caused the fault, and hence which page should be swapped in.

The design goal with ARM was to ensure that enough state is preserved inside the processor to make retrying any instruction which faults possible. In fact, load and store register instructions preserve all the CPU state when a data abort happens, so the retry is straightforward - the abort handler simply returns to the faulted instruction once the cause of the fault has been removed. This is possible for these instructions because the abort is signalled before any state has been changed. The block data transfer instructions are much harder for the abort handling software to deal with!

4.1.5 Block Data Transfer Instructions

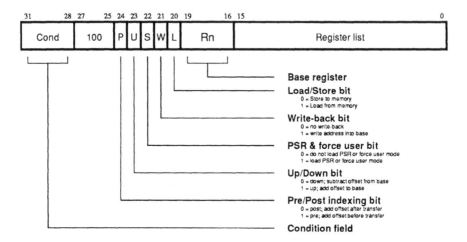

Block data transfer instructions are used to load or store any subset of the currently visible registers. They support all possible stacking modes, maintaining full or empty stacks which can grow up or down memory, and are very efficient instructions for saving or restoring context, or for moving large blocks of data around main memory.

The instruction can cause the transfer of any registers in the current bank (and non-user mode programs can also transfer to and from the user bank). The register list is a 16-bit field in the instruction, with each bit corresponding to a register. A 1 in bit 0 of the register field will cause R0 to be transferred, a 0 will cause it not to be transferred; similarly bit 1 controls the transfer of R1, and so on.

The transfer addresses are determined by the contents of the base register (Rn), the pre/post bit (P) and the up/down bit (U). The registers are transferred in the order lowest to highest, so R15 (if in the list) will always be transferred last. The lowest register also gets transferred to/from the lowest memory address.

Load Multiple Registers Implementation

The first cycle of LDM is used to calculate the address of the first word to be transferred, whilst performing a prefetch from memory (figure 100(1)). Note that the transfers always start at the lowest address and increment up memory, so the address of the first transfer may depend on the number of registers to be transferred, and a dedicated piece of logic is used to generate a count of the number of 1s in the bottom 16 bits of the instruction for this purpose.

The second cycle (figure 100(2)) fetches the first word, and performs the base modification (which may also depend on the transfer count). During the third cycle (figure 100(3)), the first word is moved to the appropriate destination register (specified by Rp, the output of a priority encoder which is used to scan through the register list in the bottom 16 bits of this instruction), while the second word is fetched from memory, and the modified base is moved to the ALU A bus input latch for holding in case it is needed to patch up after an abort.

The third cycle is repeated for subsequent fetches until the address for the last data word has been generated, though without repeating the transfer of the modified base to the ALU latch (figure 100(4)). During the transfer of the last memory word the PC is output as the address for the next cycle (figure 100(5)), then the final (internal) cycle moves the last word to its destination register (figure 100(6)) whilst supplying the PC as the address again.

The last cycle may be merged with the next instruction prefetch to form a single memory N-cycle, since the same PC value is used as the address for both cycles. The only exception is when the register list includes the PC, whereupon the next instruction prefetch will have to be a full N-cycle on its own.

The six states shown in figure 100 are not individually decoded; (3), (4) and (5) are all variations on the same decode, and in the case of the load of just one register, all three are skipped. This instruction was by far the most complex to design, since potentially it modifies all the currently visible CPU state, some of which must be preserved if recovery from a memory abort is to be possible.

If an abort occurs, the instruction continues to completion, but all register modification ceases as soon as the abort is detected. The final cycle is altered to restore the modified base register (which may have been overwritten by the load activity before the abort occurred). The base register has been preserved in the unused ALU input latch. It is restored by changing the ALU operation in the last cycle from =B to =A, and the destination for the ALU result from Rp to Rn.

If the PC is the base, write-back is disabled independently of the W bit in the instruction.

Table 20. Load Multiple memory interface signals

	Cycle	address	$\overline{b/w}$	$\overline{r/w}$	data	seq	$\overline{mreq}$	$\overline{opc}$
1 register	1	pc+8	1	0	(pc+8)	0	0	0
	2	alu	1	0	(alu)	0	1	1
	3	pc+12 pc+12	1	0	–	1	0	1
1 register	1	pc+8	1	0	(pc+8)	0	0	0
dest=pc	2	alu	1	0	pc'	0	1	1
	3	pc+12	1	0	–	0	0	1
	4	pc'	1	0	(pc')	1	0	0
	5	pc'+4 pc'+8	1	0	(pc'+4)	1	0	0
n registers	1	pc+8	1	0	(pc+8)	0	0	0
(n>1)	2	alu	1	0	(alu)	1	0	1
	.	alu+.	1	0	(alu+.)	1	0	1
	n	alu+.	1	0	(alu+.)	1	0	1
	n+1	alu+.	1	0	(alu+.)	0	1	1
	n+2	pc+12 pc+12	1	0	–	1	0	1
n registers	1	pc+8	1	0	(pc+8)	0	0	0
(n>1)	2	alu	1	0	(alu)	1	0	1
incl. pc	.	alu+.	1	0	(alu+.)	1	0	1
	n	alu+.	1	0	(alu+.)	1	0	1
	n+1	alu+.	1	0	pc'	0	1	1
	n+2	pc+12	1	0	–	0	0	1
	n+3	pc'	1	0	(pc')	1	0	0
	n+4	pc'+4 pc'+8	1	0	(pc'+4)	1	0	0

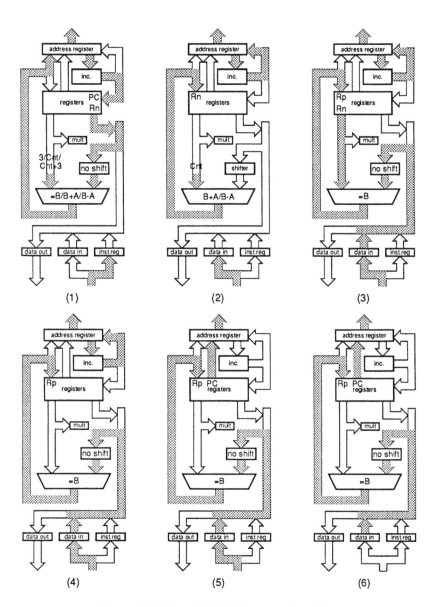

Figure 100: Load Multiple datapath activity

When the PC is in the list of registers to be loaded, and assuming that none of the transfers cause an abort to be signalled, the instructions waiting in the pipeline are marked as invalid. New instructions are fetched from the loaded PC value to replace them.

Note that the PC is always the last register to be loaded, so an abort on any of the transfers will prevent the PC from being overwritten. This is a vital feature for recovery from a page fault!

The only processor state which is guaranteed to be preserved through an abort is the PC and the base register (which may have been modified as a result of auto-indexing). This is enough to retry the instruction, though the abort handler will have to reverse any auto-indexing which has happened. Even though a load multiple instruction may auto-index the base, then overwrite it during the load, and then cause an abort on a subsequent transfer, it can be be retried successfully.

The earliest design proposal for the ARM implemented load and store multiple with an address decrementer for modes which decrement from the base address. Indeed, the first 3 micron implementation has an address incrementer/decrementer on the silicon, though it is only ever used for incrementing. The design was changed because a decrementing order would cause the PC to be overwritten first, and a subsequent transfer might then abort. Recovering from an abort when the PC has been corrupted is not possible; an additional register would be required to hold a copy of the PC in case it was needed for abort recovery. Modifying the instruction always to increment the address, and therefore to transfer the PC last, is a simpler solution (and also simplifies the memory control logic).

Store Multiple Registers Implementation

Store multiple proceeds very much as load multiple, without the final cycle. The restart problem is more straightforward here, as much less processor state is changed by the instruction.

A Note on the Block Data Transfer Instructions

The block data transfer instructions on ARM are complex by RISC standards. Considerable thought went into their design; simpler schemes such as allowing only contiguous blocks of registers were considered, but ultimately this full implementation was adopted. Two major pieces of on-chip hardware are used only by these instructions. The first is a counter circuit which gives the number of '1's in the bottom sixteen bits of the instruction. The second is

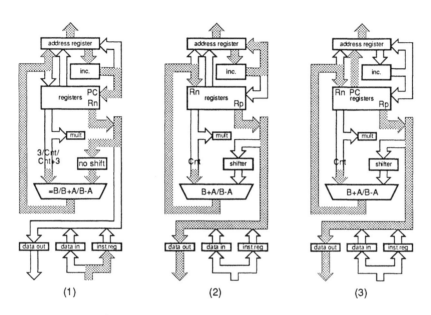

Figure 101: Store Multiple datapath activity

a sequential priority encoder which locates the least significant '1' in the same sixteen bit field, and can then mask it out in the next cycle to find the next '1', and so on.

When ARM is coupled to DRAM in the normal way (using page mode for sequential cycles), these instructions have four times the data transfer rate of the equivalent sequence of single register transfer instructions. They significantly improve the efficiency of procedure entry and return, context switching, and the movement of large blocks of data around memory.

The implementation on ARM differs from a typical microcoded implementation of such an instruction in that the main instruction decoder is put into a fixed state once the transfers are underway, and control is given over to a dedicated hardware unit which determines when to exit this state. This approach is used for all ARM instructions which do not complete in a fixed number of cycles, included in which are the multiply and coprocessor instructions. In the case of block data transfers, the loop termination condition arises when the priority encoder runs out of '1's in the bottom sixteen bits. A microcoded implementation would probably scan the sixteen bit field sequentially rather than use priority encoder hardware, and the

Table 21. Store Multiple memory interface signals

	Cycle	address	b/w	r̄/w̄	data	seq	m̄r̄ēq̄	ōp̄c̄
1 register	1	pc+8	1	0	(pc+8)	0	0	0
	2	alu	1	1	Ra	0	0	1
n registers	1	pc+8	1	0	(pc+8)	0	0	0
(n>1)	2	alu	1	1	Ra	1	0	1
	.	alu+.	1	1	R.	1	0	1
	n	alu+.	1	1	R.	1	0	1
	n+1	alu+.	1	1	R.	0	0	1

microcode program would terminate naturally when the scan had completed. This would always take sixteen micro-cycles, whereas ARM requires only a cycle for each '1'.

4.1.6 Software Interrupt Instruction

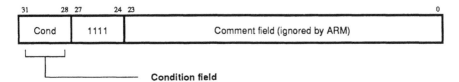

The software interrupt instruction is used to enter supervisor mode in a controlled manner. The instruction causes the software interrupt trap to be taken, which forces the processor into supervisor mode and causes the PC to jump to the value 08H. If this address is suitably protected from modification by the user (by external memory management hardware), a fully protected operating system may be constructed.

The PC and PSR are saved in R14_svc upon entering the software interrupt trap, with the PC adjusted to point to the word after the SWI instruction. MOVS R15,R14_svc will return to the user program, restore the user PSR and return the processor to user mode. Note that the link mechanism is not re-entrant, so if the supervisor code uses subroutines or software interrupts within itself it must first save a copy of the return address.

The bottom 24 bits of the instruction are ignored by ARM, and may be used to communicate information to the supervisor code. For instance, the supervisor may look at this field (via R14) and use it to index into an array of entry points for routines which perform the various supervisor functions.

Software Interrupt and Exception Entry Implementation

Exceptions (and software interrupts) force the PC to a particular value and refill the instruction pipeline from there. During the first cycle the forced address is constructed, and a mode change may take place. The return address is moved to the supervisor mode copy of register 14.

During the second cycle the return address is modified to facilitate return, exactly as was described for the branch with link instruction. It gives the correct return address for software interrupts, but is out by one word for most other exceptions, and out by two words for exceptions that return to retry the faulted instruction. The third cycle is required only to complete the refilling of the instruction pipeline.

In the table of CPU pin activity, 'pc' is the address of the SWI instruction; for interrupts and reset 'pc' is the address of the instruction following the last

Table 22. Software Interrupt and Exception entry memory interface signals

Cycle	address	b/w	r/w	data	seq	mreq	opc	trans
1	pc+8	1	0	(pc+8)	0	0	0	1
2	Xn	1	0	(Xn)	1	0	0	1
3	Xn+4	1	0	(Xn+4)	1	0	0	1
	Xn+8							

one to be executed before entering the exception; for prefetch abort 'pc' is the address of the aborting instruction; for data abort 'pc' is the address of the instruction following the one which attempted the aborted data transfer. 'Xn' is the appropriate trap address.

The following instructions were not implemented on the first ARM chip, but were added to the second (2 micron) version of the ARM:

4.1.7 Multiply and Multiply-Accumulate Instructions

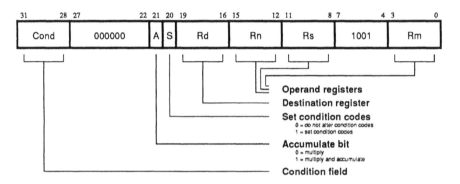

The multiply and multiply-accumulate instructions use a 2-bit Booth's algorithm to perform integer multiplication. They give the least significant 32 bits of the product of two 32-bit operands, and may be used to synthesize higher precision multiplications.

The multiply form of the instruction gives Rd:=Rm*Rs. Rn is ignored. The multiply-accumulate form gives Rd:=Rm*Rs+Rn, which can save an explicit ADD instruction in some circumstances. Both forms of the instruction work on operands which may be considered as signed (2's complement) or unsigned integers.

The multiply instruction is a complex instruction requiring its own state machine controller, but very little additional logic. It was added to the second version of the chip for two reasons. Firstly, although multiply is not too inefficient when done in-line with standard data processing instructions, it does require several instructions. Compiler writers had tended to avoid the code length penalty of the in-line approach, and the consequent problems of register allocation, by using multiply subroutines. The most flexible implementations called subroutines which interpreted fields at the location of the caller to determine which registers to use; these were horribly slow. The second reason was the observation that ARMs were likely to be used in applications which required some signal processing, for instance in the Acorn chip set the CPU generates audio samples by real-time digital signal processing (DSP). A small improvement in multiply performance can make a lot of difference to real-time DSP. The on-chip multiply gave a four-fold improvement in multiply time, and helped both of these areas. It had very low hardware cost, and all the additional components were attached off critical timing paths.

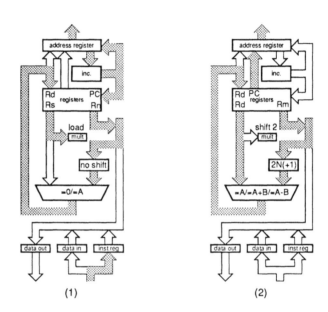

Figure 102: Multiply datapath activity

Multiply and Multiply-Accumulate Implementation

The multiply instructions make use of special hardware with which is implemented a 2-bit Booth's algorithm with early termination (see table 23). During the first cycle the accumulate register is brought to the ALU, which either transmits it or produces zero (according to whether the instruction is MLA or MUL) to initialize the destination register. During the same cycle one of the operands is loaded into the Booth's shifter via the A bus.

The datapath then cycles, adding the second operand to, subtracting it from, or just transmitting, the result register. The second operand is shifted in the Nth cycle (where we start with N=0, like all computer scientists!) by 2N or 2N+1 bits, under control of the Booth's logic. The first operand is shifted right 2 bits per cycle, and when it is zero the instruction terminates (possibly

Table 23. The 2-bit Booth's algorithm used on ARM2

```
(* 2-bit Booth's algorithm:                              *)
(*              result:= op1*op2 {possibly + op3}  *)
(*              uses A + 3B = A - B + 4B;                *)
(*              4B is borrowed from the next cycle.*)


borrow := FALSE; N:= 0;

IF accumulate THEN result:=op3 ELSE result:=0 END;

WHILE ((op2 # 0) OR borrow) AND N<16 DO
  IF NOT borrow THEN
    CASE op2 MOD 3 OF
       0:  result:= result;
     | 1:  result:= result + op1*(2^(2*N))   ;
     | 2:  result:= result - op1*(2^(2*N+1));
           borrow := TRUE ;
     | 3:  result:= result - op1*(2^(2*N))   ;
           borrow := TRUE ;
    END; (* case *)
  ELSE                      (* borrow is TRUE *)
    CASE op2 MOD 3 OF
       0:  result:= result + op1*(2^(2*N))   ;
           borrow := FALSE;
     | 1:  result:= result + op1*(2^(2*N+1));
           borrow := FALSE;
     | 2:  result:= result - op1*(2^(2*N))   ;
     | 3:  result:= result;
    END; (* case *)
  END; (* if *)
  op2:= op2 DIV 4; N:= N+1;   (* op2 >> 2 bits *)
END; (* while *)
```

after an additional cycle to clear a pending borrow). A borrow from the 15th cycle is discarded, as it does not affect the least significant 32 bits. (These instructions only generate the bottom 32 bits of the result.)

All cycles except the first are internal.

If the destination is the PC, all writing to it is prevented, and the instruction is useless but harmless. The instruction will proceed as normal except that the PC will be unaffected. (If the S bit is set the PSR flags will be corrupted in a meaningless way.)

In the table of CPU pin activity (table 24), 'm' is the number of cycles required by the Booth's algorithm (see the section on instruction speeds).

Table 24. Multiply memory interface signals

	Cycle	address	b/w	r̄/w̄	data	seq	m̄r̄ēq̄	ōp̄c̄
(Rs)=0,1	1	pc+8	1	0	(pc+8)	0	1	0
	2	pc+12	1	0	–	1	0	1
		pc+12						
(Rs)>1	1	pc+8	1	0	(pc+8)	0	1	0
	2	pc+12	1	0	–	0	1	1
	.	pc+12	1	0	–	0	1	1
	m	pc+12	1	0	–	0	1	1
	m+1	pc+12	1	0	–	1	0	1
		pc+12						

Like the block data transfer instructions, multiply puts the instruction decoder into a fixed state and control is taken over by dedicated hardware. In this case, the termination condition which brings the processor out of the loop is when sixteen cycles have been performed, or there are no more ones in the two-bit shifter and no pending borrow. The state machine which is used to implement the Booth's algorithm is completely separate from the central instruction decoder, and is not used at any time other than during multiplies.

The 32-bit result is adequate for many DSP applications, and it is clearly sufficient for calculating the address of an element of a multi-dimensional array (since all addresses are less than 32 bits), but it is not adequate for many floating-point applications. A 64-bit result may be obtained by splitting each of the 32-bit operands into two 16-bit halves, performing the four component multiplications, and realigning and recombining the results. The component multiplications will all benefit from the early termination condition, since the operands are restricted to 16 bits, so only 32 multiply cycles are needed to produce all the intermediate products (possibly with extra cycles for borrows), and a few data operations are needed to combine them. The instruction set examples later in this chapter include such a routine.

Other new instructions were added to the second ARM to allow the connection of hardware coprocessors, in particular floating-point hardware:

4.1.8 Coprocessor Data Operation Instructions

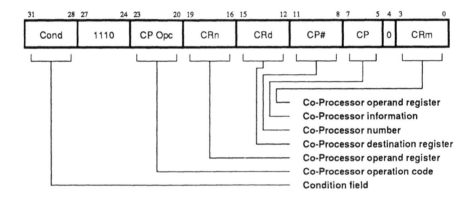

This class of instruction is used to tell a coprocessor to perform some internal operation. No result is communicated back to ARM or to memory, and ARM will not wait for the operation to complete. The coprocessor could contain a queue of such instructions awaiting execution, and their execution can overlap other ARM activity allowing the coprocessor and ARM to perform independent tasks in parallel.

Only bit 4 and bits 24 to 31 are significant to ARM; the remaining bits are significant only to coprocessors. The above field names are used by convention, and particular coprocessors may redefine the use of all fields except CP# as appropriate. The CP# field is used to contain a unique identifying number (in the range 0 to 15) for each coprocessor. A coprocessor will ignore any instruction which does not contain its number in the CP# field.

The conventional interpretation of the instruction is that the coprocessor should perform an operation specified in the CP Opc field (and possibly in the CP field) on the contents of CRn and CRm, and place the result in CRd.

Coprocessor Data Operation Implementation

All coprocessor instructions (and undefined instructions) are indicated as such by ARM taking $\overline{\text{CPI}}$ (coprocessor instruction) low. The coprocessor (if present) responds by taking CPA (coprocessor absent) low, and, when it is ready to undertake the requested activity, it takes **CPB** (coprocessor busy) low.

A coprocessor data operation is a request from ARM for the coprocessor to initiate some action. The action need not be completed for some time, but the coprocessor must commit to doing it before pulling **CPB** low.

If the coprocessor can never do the requested task, it should leave **CPA** and **CPB** to float high. If it can do the task, but can't commit right now, it should pull **CPA** low but leave **CPB** high until it can commit. ARM will busy-wait until **CPB** goes low.

Table 25. Coprocessor Data Operation memory interface signals

Cycle	address	b/w	$\overline{r/w}$	data	seq	$\overline{mreq}$	$\overline{opc}$	$\overline{cpi}$	$\overline{cpa}$	$\overline{cpb}$
ready 1	pc+8	1	0	(pc+8)	1	0	0	0	0	0
	pc+12									
not 1	pc+8	1	0	(pc+8)	0	1	0	0	0	1
ready 2	pc+8	1	0	–	0	1	1	0	0	1
.	pc+8	1	0	–	0	1	1	0	0	1
n	pc+8	1	0	–	0	0	1	0	0	0
	pc+12									

The most complex part of the coprocessor interface is the interruptible busy-wait state. If an interrupt happens while the coprocessor is busy, the return from interrupt must retry the coprocessor instruction, whereas if the interrupt arrives just as the coprocessor commits to performing the requested operation, the return must not retry the instruction.

All standard ARM instructions begin with an instruction prefetch during the first cycle, and copy the incremented address (which must itself be the current PC) back into the PC. The coprocessor instructions perform the same prefetch during the first cycle, but make the copy of the incremented PC conditional upon the coprocessor being ready. If the coprocessor is busy, the

PC is not incremented, and ARM goes into an internal cyclical state waiting for it to go not-busy. If an interrupt arrives while the coprocessor is still busy, the unincremented PC gets copied into the interrupt link register. This is exactly as though the coprocessor instruction never started, so the return will go back to retry the coprocessor instruction. If the coprocessor goes not-busy before the interrupt arrives, the PC is incremented, and a subsequent interrupt will save the incremented value to the link register and return to the instruction following the coprocessor instruction.

4.1.9 Coprocessor Data Transfer Instructions

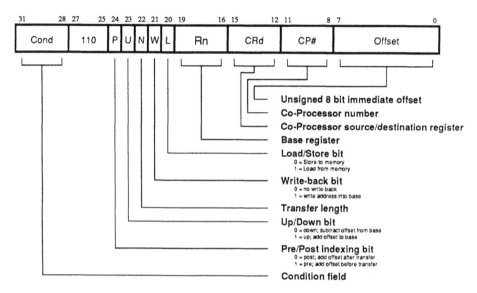

This class of instruction is used to transfer one or more words of data between the coprocessor and main memory. ARM is responsible for supplying the memory address, and the coprocessor supplies or accepts the data and controls the number of words transferred.

The CP# field is used to identify the coprocessor which is required to supply or accept the data, and a coprocessor will only respond if its number matches the contents of this field.

The CRd field and the N bit contain information for the coprocessor which may be interpreted in different ways by different coprocessors, but by convention CRd is the register to be transferred (or the first register where more than one is to be transferred), and the N bit is used to choose one of two transfer length options. For instance N=0 could select the transfer of a single register, and N=1 could select the transfer of all the registers for context switching.

ARM is responsible for providing the address used by the memory system for the transfer, and the addressing modes available are a subset of those used in single data transfer instructions. Note, however, that the immediate offsets use 8 bits and specify word offsets here, whereas they use 12 bits and specify byte offsets for single register transfers.

An 8-bit unsigned immediate offset is scaled to words (ie shifted left 2 bits) and added to (U=1) or subtracted from (U=0) a base register (Rn), either before (P=1) or after (P=0) the base is used as the transfer address. The modified base value may be overwritten back into the base register (if W=1), or the old value of the base may be preserved (W=0).

The value of the base register, modified by the offset in a pre-indexed instruction, is used as the address for the transfer of the first word. The second word (if more than one is transferred) will go to or come from an address one word (4 bytes) higher than the first transfer, and the address will continue to be incremented by one word for each subsequent transfer.

Coprocessor Data Transfer (to Coprocessor) Implementation

Here the coprocessor should commit to the transfer only when it is ready to accept the data. When **CPB** goes low, ARM will produce addresses and expect the coprocessor to take the data at sequential cycle rates. The coprocessor is responsible for determining the number of words to be transferred, and indicates the last transfer cycle by allowing **CPA** and **CPB** to float high.

ARM spends the first cycle (and any busy-wait cycles) generating the transfer address, and performs the write-back of the address base during the transfer cycles.

Coprocessor Data Transfer (from Coprocessor) Implementation

The ARM controls these instructions exactly as for memory to coprocessor transfers, with the one exception that the $\overline{R/W}$ line is inverted during the transfer cycles.

Table 26. Data Transfer to coprocessor memory interface signals

Cycle	address	$\overline{b}/\overline{w}$	r/w	data	seq	$\overline{mreq}$	$\overline{opc}$	$\overline{cpi}$	cpa	cpb
1 reg 1	pc+8	1	0	(pc+8)	0	0	0	0	0	0
ready 2	alu pc+12	1	0	(alu)	0	0	1	1	1	1
1 reg 1	pc+8	1	0	(pc+8)	0	1	0	0	0	1
not 2	pc+8	1	0	–	0	1	1	0	0	1
ready .	pc+8	1	0	–	0	1	1	0	0	1
n	pc+8	1	0	–	0	0	1	0	0	0
n+1	alu pc+12	1	0	(alu)	0	0	1	1	1	1
n reg 1	pc+8	1	0	(pc+8)	0	0	0	0	0	0
(n>1) 2	alu	1	0	(alu)	1	0	1	1	0	0
ready .	alu+.	1	0	(alu+.)	1	0	1	1	0	0
n	alu+.	1	0	(alu+.)	1	0	1	1	0	0
n+1	alu+. pc+12	1	0	(alu+.)	0	0	1	1	1	1
m reg 1	pc+8	1	0	(pc+8)	0	1	0	0	0	1
(m>1) 2	pc+8	1	0	–	0	1	1	0	0	1
not .	pc+8	1	0	–	0	1	1	0	0	1
ready n	pc+8	1	0	–	0	0	1	0	0	0
n+1	alu	1	0	(alu)	1	0	1	1	0	0
ready .	alu+.	1	0	(alu+.)	1	0	1	1	0	0
n+m	alu+.	1	0	(alu+.)	1	0	1	1	0	0
n+m+1	alu+. pc+12	1	0	(alu+.)	0	0	1	1	1	1

Table 27. Data Transfer from coprocessor memory interface signals

Cycle	address	$\overline{b/w}$	$\overline{r/w}$	data	seq	$\overline{mreq}$	$\overline{opc}$	$\overline{cpi}$	cpa	cpb
1 reg 1	pc+8	1	0	(pc+8)	0	0	0	0	0	0
ready 2	alu pc+12	1	1	CPdata	0	0	1	1	1	1
1 reg 1	pc+8	1	0	(pc+8)	0	1	0	0	0	1
not 2	pc+8	1	0	–	0	1	1	0	0	1
ready .	pc+8	1	0	–	0	1	1	0	0	1
n	pc+8	1	0	–	0	0	1	0	0	0
n+1	alu pc+12	1	1	CPdata	0	0	1	1	1	1
n reg 1	pc+8	1	0	(pc+8)	0	0	0	0	0	0
(n>1) 2	alu	1	1	CPdata	1	0	1	1	0	0
ready .	alu+.	1	1	CPdata	1	0	1	1	0	0
n	alu+.	1	1	CPdata	1	0	1	1	0	0
n+1	alu+. pc+12	1	1	CPdata	0	0	1	1	1	1
m reg 1	pc+8	1	0	(pc+8)	0	1	0	0	0	1
(m>1) 2	pc+8	1	0	–	0	1	1	0	0	1
not .	pc+8	1	0	–	0	1	1	0	0	1
ready n	pc+8	1	0	–	0	0	1	0	0	0
n+1	alu	1	1	CPdata	1	0	1	1	0	0
ready .	alu+.	1	1	CPdata	1	0	1	1	0	0
n+m	alu+.	1	1	CPdata	1	0	1	1	0	0
n+m+1	alu+. pc+12	1	1	CPdata	0	0	1	1	1	1

A Note on the Coprocessor Data Transfer Instructions

In many ways these instructions resemble the ARM data transfer instructions. The initial address calculation is very similar to the single register transfer instructions, and subsequent address incrementing activity is similar to the block data transfers. The abort handling is much simpler than in the case of load multiple registers, since the only change in ARM state is the auto-indexing of the base register.

This instruction class represents an extreme case of passing control to dedicated hardware to determine when a frozen instruction state should terminate. Here, the dedicated logic is in the coprocessor, and not in ARM. ARM will continue passively incrementing the address for ever if it is not told to stop. By convention, a coprocessor is not allowed to continue the transfer beyond sixteen words (so that interrupt latency cannot be impacted by these instructions), but there is no hardware enforcement of this rule within ARM itself.

4.1.10 Coprocessor Register Transfer Instructions

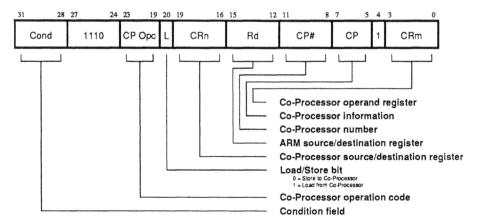

This class of instruction is used to communicate information directly between ARM and a coprocessor. An example of an MCR (Move from Coprocessor to ARM) instruction would be a FIX of a floating-point value held in a coprocessor, where the floating-point number is converted into a 32-bit integer within the coprocessor, and the result is then transferred to an ARM register. A FLOAT of a 32-bit value in an ARM register into a floating-point value within the coprocessor illustrates a possible use of MRC (Move from ARM to coprocessor).

An important use of this instruction is to communicate control information directly from the coprocessor into the ARM PSR flags. As an example, the result of a comparison of two floating-point values within a coprocessor can be moved to the PSR to control the subsequent flow of execution.

The CP# field is used, as for all coprocessor instructions, to specify which coprocessor is being called upon to respond.

The CP Opc, CRn, CP and CRm fields are used only by the coprocessor, and the interpretation presented here is derived from convention only. Other interpretations are allowed where the coprocessor functionality is incompatible with this one. The conventional interpretation is that the CP Opc and CP fields specify the operation the coprocessor is required to perform, CRn is the coprocessor register which is the source or destination of the transferred information, and CRm is a second coprocessor register which may be involved in some way which depends on the particular operation specified.

Coprocessor Register Transfer (Load from Coprocessor) Implementation

Here the busy-wait cycles are much as they are for the previous coprocessor instructions, but the transfer is limited to one data word, and ARM puts the transferred word into the destination register in the third cycle. The third cycle may be merged with the following prefetch cycle into one memory N-cycle as with all ARM register load instructions.

Table 28. Register Transfer from coprocessor memory interface signals

	Cycle	address	b/w	r/w	data	seq	mreq	opc	cpi	cpa	cpb
ready	1	pc+8	1	0	(pc+8)	1	1	0	0	0	0
	2	pc+12	1	0	CPdata	0	1	1	1	1	1
	3	pc+12	1	0	–	1	0	1	1	–	–
		pc+12									
not	1	pc+8	1	0	(pc+8)	0	1	0	0	0	1
ready	2	pc+8	1	0	–	0	1	1	0	0	1
	.	pc+8	1	0	–	0	1	1	0	0	1
	n	pc+8	1	0	–	1	1	1	0	0	0
	n+1	pc+12	1	0	CPdata	0	1	1	1	1	1
	n+2	pc+12	1	0	–	1	0	1	1	–	–
		pc+12									

This instruction may require internal coprocessor activity before the 32-bit value is ready for transfer, for instance a floating-point FIX operation must perform the floating-point to fixed-point conversion. During the conversion, the coprocessor must cause ARM to wait by keeping the busy signal active. While the busy line is active, ARM may be interrupted. The pre-transfer activity in the coprocessor must therefore be restartable. The coprocessor may keep the intermediate result in case the next operation it is asked to do is the retry (which will usually be the case), or it may start from the beginning each time.

Coprocessor Register Transfer (Store to Coprocessor) Implementation

This is similar to the load from coprocessor; it differs in that the last cycle is omitted, and the read/write signal is inverted for the transfer cycle.

Table 29. Register Transfer to coprocessor memory interface signals

	Cycle	address	b/w	$\overline{\text{r/w}}$	data	seq	$\overline{\text{mreq}}$	$\overline{\text{opc}}$	$\overline{\text{cpi}}$	cpa	cpb
ready	1	pc+8	1	0	(pc+8)	1	1	0	0	0	0
	2	pc+12	1	1	Rd	0	0	1	1	1	1
		pc+12									
not	1	pc+8	1	0	(pc+8)	0	1	0	0	0	1
ready	2	pc+8	1	0	–	0	1	1	0	0	1
	.	pc+8	1	0	–	0	1	1	0	0	1
	n	pc+8	1	0	–	1	1	1	0	0	0
	n+1	pc+12	1	1	Rd	0	0	1	1	1	1
		pc+12									

Any operation performed by the coprocessor is likely to follow the transfer, so the handshake will usually be more straightforward than for transfers in the other direction. The coprocessor just has to be ready to latch the transferred word before it indicates readiness to proceed.

4.1.11 Undefined Instructions and Coprocessor Absent

When a coprocessor detects a coprocessor instruction which it cannot perform (which must include all undefined instructions), it must not drive **CPA** or **CPB**. These will float high, causing the undefined instruction trap to be taken. The entry sequence is similar to the standard exception entry sequence, with the addition of a cycle at the beginning which is required for ARM to establish via the coprocessor handshake that no coprocessor is present that can perform the instruction.

Table 30. Undefined instruction memory interface signals

Cycle	address	b/w	r/w	data	seq	mreq	opc	cpi	cpa	cpb
1	pc+8	1	0	(pc+8)	0	1	0	0	1	1
2	pc+8	1	0	–	0	0	0	1	1	1
3	Xn	1	0	(Xn)	1	0	0	1	1	1
4	Xn+4	1	0	(Xn+4)	1	0	0	1	1	1
	Xn+8									

4.1.12 Unexecuted Instructions

Any instruction whose condition code is not met will fail to execute. It will add one cycle to the execution time of the code segment in which it is embedded.

The condition code evaluation is performed concurrently with the operand fetch phase of the first cycle, and the failure of the condition test causes all result writing to be disabled and subsequent cycles to be skipped; instead the next instruction starts on the next cycle.

Table 31. Unexecuted instruction memory interface signals

Cycle	address	$\overline{b/w}$	$\overline{r/w}$	data	seq	$\overline{mreq}$	$\overline{opc}$
1	pc+8 pc+12	1	0	(pc+8)	1	0	0

4.2 INSTRUCTION SET EXAMPLES

The following examples show ways in which the basic ARM instructions can be combined to give efficient code. None of these methods saves a great deal of execution time (although they may save some), mostly they just save code.

(i) Using the conditional instructions:

(1) using conditionals for logical OR

```
CMP     Rn,#p  ;IF Rn=p OR Rm=q THEN GOTO Label
BEQ     Label
CMP     Rm,#q
BEQ     Label
```

can be replaced by

```
CMP     Rn,#p
CMPNE   Rm,#q  ;if not satisified, other test
BEQ     Label
```

(2) absolute value

```
TEQ     Rn,#0           ;test sign
RSBMI   Rn,Rn,#0        ;2's complement if -ve
```

(3) multiplication by 4, 5 or 6 (run time)

```
MOV     Rc,Ra,LSL #2  ;multiply by 4
CMP     Rb,#5         ;test value
ADDCS   Rc,Rc,Ra      ;complete multiply by 5
ADDHI   Rc,Rc,Ra      ;complete multiply by 6
```

(4) combining discrete and range tests

```
TEQ     Rc,#127       ;discrete test
CMPNE   Rc,#" "-1     ;range test
MOVLS   Rc,#"."       ;IF Rc<=" " OR Rc=CHR$127
                      ;THEN Rc:="."
```

(5)   division and remainder

```
;enter with numbers in Ra and Rb
       MOV    Rcnt,#1          ;bit to control div.
Div1 CMP     Rb,#&80000000 ;move Rb until > Ra
       CMPCC  Rb,Ra
       MOVCC  Rb,Rb,ASL #1
       MOVCC  Rcnt,Rcnt,ASL #1
       BCC    Div1
       MOV    Rc,#0
Div2 CMP     Ra,Rb                ;test subtraction
       SUBCS  Ra,Ra,Rb             ;subtract if ok
       ADDCS  Rc,Rc,Rcnt           ;put bit into result
       MOVS   Rcnt,Rcnt,LSR #1;shift control bit
       MOVNE  Rb,Rb,LSR #1    ;halve unless done
       BNE    Div2
;divide result in Rc
;remainder in Ra
```

(ii)   Pseudo random binary sequence generator:

It is often necessary to generate (pseudo-) random numbers, and the most efficient algorithms are based on a shift register with exclusive-or feedback, rather like a cyclic redundancy check generator. Unfortunately, the sequence of a 32-bit generator needs more than one feedback tap to be maximal length (ie $2^{32}-1$ cycles before repetition). Therefore BBC BASIC uses a 33-bit register with taps at bits 33 and 20. The basic algorithm is newbit:=bit33 eor bit20, shift left the 33-bit number, and put in newbit at the bottom. Then do this for all the newbits needed, ie 32 times. Luckily, this can all be done in 5 instructions:

```
;enter with seed in Ra (32 bits), Rb (lsb)
;uses Rc
TST    Rb,Rb,LSR #1    ;top bit into carry
MOVS   Rc,Ra,RRX       ;33-bit rotate right
ADC    Rb,Rb,Rb        ;carry into lsb of Rb
EOR    Rc,Rc,Ra,LSL#12 ;top exclusive-ors
EOR    Ra,Rc,Rc,LSR#20 ;bottom exclusive-ors
;new seed in Ra, Rb as before
```

(iii) Multiplication by constant using the barrel shifter:

(1) Multiplication by 2^n (1,2,4,8,16,32..)

```
MOV    Ra,Ra,LSL #n
```

(2) Multiplication by 2^n+1 (3,5,9,17..)

```
ADD    Ra,Ra,Ra,LSL #n
```

(3) Multiplication by 2^n-1 (3,7,15..)

```
RSB    Ra,Ra,Ra,LSL #n
```

(4) Multiplication by 6

```
ADD    Ra,Ra,Ra,LSL #1 ;multiply by 3
MOV    Ra,Ra,LSL #1    ;and then by 2
```

(5) Multiply by 10 and add in extra number

```
ADD    Ra,Ra,Ra,LSL #2 ;multiply by 5
ADD    Ra,Rc,Ra,LSL #1 ;times 2 plus extra no.
```

(6) General recursive method for Rb := Ra*C, C a constant:

(a) If C even, say $C = 2^n*D$, D odd:

```
D=1:  MOV    Rb,Ra,LSL #n
D<>1: {Rb := Ra*D}
      MOV    Rb,Rb,LSL #n
```

(b) If C MOD 4 = 1, say $C = 2^n*D+1$, D odd, n>1:

```
D=1:  ADD    Rb,Ra,Ra,LSL #n
D<>1: {Rb := Ra*D}
      ADD    Rb,Ra,Rb,LSL #n
```

(c) If C MOD 4 = 3, say $C = 2^n*D-1$, D odd, n>1:

```
D=1:  RSB    Rb,Ra,Ra,LSL #n
D<>1: {Rb := Ra*D}
      RSB    Rb,Ra,Rb,LSL #n
```

This algorithm is not quite optimal, but it is not far off. An example of its sub-optimality is multiply by 45, which gives:

```
RSB    Rb,Ra,Ra,LSL #2 ;multiply by 3
RSB    Rb,Ra,Rb,LSL #2 ;multiply by 4*3-1 = 11
ADD    Rb,Ra,Rb,LSL #2 ;multiply by 4*11+1 = 45
```

rather than by:

```
ADD    Rb,Ra,Ra,LSL #3 ;multiply by 9
ADD    Rb,Rb,Rb,LSL #2 ;multiply by 5*9 = 45
```

(iv)   Loading a word from an unknown alignment:

```
       ;enter with address in Ra (32 bits)
       ;uses Rb, Rc; result in Rd.
       ;Note d must be less than c e.g. 0,1
BIC    Rb,Ra,#3          ;get word aligned address
LDMIA  Rb,{Rd,Rc}        ;get 64 bits containing data
AND    Rb,Ra,#3          ;correction factor in bytes
MOVS   Rb,Rb,LSL #3      ;now in bits; test if aligned
MOVNE  Rd,Rd,LSR Rb      ;bottom of result word
                         ;(if not aligned)
RSBNE  Rb,Rb,#32      ;get other shift amount
ORRNE  Rd,Rd,Rc,LSL Rb ;combine two halves
```

(v)   Sign/zero extension of a half word:

```
MOV    Ra,Ra,LSL #16     ;move to top
MOV    Ra,Ra,LSR #16     ;and back to bottom
                         ;use ASR to get
                         ;sign extended version
```

(vi) Return setting condition codes:

```
      BICS    PC,R14,#CFLAG  ;returns clearing C flag
                             ;from link register
      ORRCCS PC,R14,#CFLAG  ;conditionally returns
                             ;setting C flag

      ; This code should only be used in User mode
      ; since it will reset the interrupt enable flags
      ; to their value when R14 was set up.
      ; Generally applies to non-User mode programming
      ; e.g. MOVS PC,R14 - MOV PC,R14 is safer!
```

(vii) Extended precision multiply:

The multiply instruction may be used to synthesize higher precision multiplications, for instance to multiply two 32-bit integers to generate a 64-bit result:

```
mul64
      MOV    a1,A,LSR #16   ; a1:= top half of A
      MOV    D,B,LSR #16    ; D := top half of B
      BIC    A,A,a1,LSL #16 ; A := bottom half of A
      BIC    B,B,D,LSL #16  ; B := bottom half of B
      MUL    C,A,B          ; low section of result
      MUL    B,a1,B         ; ) middle sections
      MUL    A,D,A          ; )      of result
      MUL    D,a1,D         ; high section of result
      ADDS   A,B,A          ; add middle sections
             ; (couldn't use MLA as we need C correct)
      ADDCS D,D,#&10000     ; carry from above add
      ADDS  C,C,A,LSL #16   ; C is now bottom 32 bits
      ADC   D,D,A,LSR #16   ; D is top 32 bits
```

(A and B are registers containing the two 32-bit integer operands; C and D are registers for the 64-bit result; a1 is a temporary register. A and B are overwritten during the multiply)

4.3 INSTRUCTION SPEEDS

Due to the pipelined architecture of the CPU, instructions overlap considerably. In a typical cycle one instruction may be using the data path while the next is being decoded and the one after that is being fetched. For this reason the following table presents the incremental number of cycles required by an instruction, rather than the total number of cycles for which the instruction uses part of the processor. Elapsed time (in cycles) for a routine may be calculated from these figures.

If the condition specified in the instruction condition field is met the instructions take:

```
Data Proc   1 S                     + 1 S      for SHIFT(Rs)
                                    + 1 S + 1 N if Rd=R15
LDR         1 S + 1 N + 1 I   + 1 S + 1 N if Rd=R15
STR               2 N
LDM         n S + 1 N + 1 I   + 1 S + 1 N if {..,R15}
STM     (n-1)S + 2 N
B,BL        2 S + 1 N
SWI,trap    2 S + 1 N
MUL,MLA     1 S        + m I
CDP         1 S        + b I
LDC,STC (n-1)S + 2 N + b I
MRC         1 S        + b I + 1 C
MCR         1 S   + (b+1)I + 1 C
```

If the condition is not met all instructions take one S-cycle.

n is the number of words transferred.

m is the number of cycles required by the multiply algorithm, which is determined by the contents of Rs. Multiplication by any number between $2^{(2m-3)}$ and $2^{(2m-1)}-1$ inclusive takes m cycles for m>1. Multiplication by 0 or 1 takes 1 cycle. If the top bit of Rs is set, the multiply will take 16 cycles.

b is the number of cycles spent in the coprocessor busy-wait loop.

The cycle types (N, S, I and C) are defined in the memory interface section.

4.4 INSTRUCTION USAGE

The ARM instruction set has many features in common with other RISC instruction sets, for instance the load/store architecture, and three register addressing for data processing instructions. It also has some unusual characteristics, most notably the conditional skipping of all instructions. It is of interest, therefore, to analyse the frequency of use of the various features. Table 32 presents some results for dynamic instruction usage, based on two programs. The first set are for the compilation of a small C program, which takes 20 million processor cycles and 10 million instructions. The compiler itself is a compiled C program, so this may be typical for a C program which uses large data structures. The second set of results are for 2.5 million instructions executed by a BASIC interpreter, which represents very carefully hand-optimized assembly code.

The following observations may be made from these results:

- The condition field is frequently useful even to the compiler. 8% of the C instructions were conditional non-branches, as were 15% of the BASIC instructions. Without the condition field, these instructions would require branches to bypass them.

- BASIC combines a shift with a non-trivial ALU operation (ie not a simple move) on at least 9% of all instructions. The compiler is less successful at exploiting these compound data operations.

- The multiple register transfer instructions have low frequency, but transfer about as many registers in total as the single register transfer instructions. The compiler transfers more registers at a time because its procedure entry and exit sequences always save several registers (except when entering or exiting leaf procedures, when no registers are saved).

- Stores with base plus index addressing are rare, at around 1%. These figures would support the asymmetrical decision taken in the design of some RISC architectures to support base plus index addressing on loads only. This also allows single cycle stores with a register file with only two read buses. Since ARM uses two-cycle stores, base plus index addressing is easily incorporated, and uses the same mechanism as loads with base plus index addressing.

- The auto-indexing modes seem useful, and since they make use of the otherwise useless cycle when data is being transferred to or from memory, their inclusion is justified.

Table 32. ARM Instruction Usage

instruction type	C compilation	BASIC execution
unconditional	77 %	65 %
condition passed	10 %	14 %
condition failed	13 %	21 %
branches taken	12 %	8 %
conditionally	7 %	4 %
with link	3 %	4 %
not taken	8 %	16 %
data ops total	43 %	53 %
using ALU	30 %	39 %
using shift	10 %	23 %
load multiple	3 %	5 %
ave no. of regs	7	2
store multiple	3 %	2 %
ave no. of regs	7	2
load base + offset	19 %	1 %
base + index	2 %	7 %
store base + offset	4 %	2 %
base + index	1 %	1 %
l/s auto-indexing	3 %	5 %
data words accessed per 100 instructions	69	27

(Source: Wills, 1988)

- The comparison of data traffic at least in part reflects the different tasks the two programs are undertaking. The C compiler is building and referencing large data structures, whereas the BASIC interpreter is performing a computationally-intensive benchmark task.

Some of the ARM instructions are quite complex; the compound shift plus ALU operations, and the multiple register transfer instructions, for example. These factors confuse the 'cycles per instruction' calculations favoured by some RISC analysts. In table 33, the effects of simplifying the instruction set are postulated (in terms of the additional instructions required).

Table 33. The Effect of Modifying the ARM Instruction Set

instruction type	C compilation	BASIC execution
ARM instructions (cycles/instruction	200 cycles 100 2	150 cycles 100 1.5)
without condition	+4 to +8	+8 to +15
without multiple transfer	+36	+7
without ALU + shift	-	+9
simple instructions cycles/instruction	140 1.4	125 1.2

The C compiler program takes about 200 cycles for 100 instructions, whereas the BASIC interpreter takes 150 (it takes fewer because of the reduced data traffic). The cycles per ARM instruction are 2 and 1.5 respectively. If the condition field were removed (apart from branches), branches would have to be added. The cost would be less than one branch per conditional instruction, since sequences of two or three instructions with the same condition are quite common.

If the multiple transfer instructions were not available, a single transfer instruction would be needed for each register moved. If the compound data operation were not available, such an operation would be split into two instructions. Hence the same work might require 150 or 140 'simple' instructions to perform the work of 100 ARM instructions, and the number of ARM cycles per simple instruction is quite close to 1 in both cases.

The cost of including the more complex instructions is an increase in the basic datapath cycle time, most obviously for the compound instructions which require the barrel shifter to operate in series with the ALU. But since the ARM cycle time is normally locked to DRAM speeds, it is usually running below its maximum clock rate anyway. In such circumstances it makes sense to allow as much work as possible to be performed in a cycle, and to design the instruction set to allow access to the full functionality of the datapath.

4.5 DESIGN METHODOLOGY

Up to this point the instruction set design and datapath activity diagrams are largely a paper exercise. The next step is to build a computer model which can be used to test the details of the design. The computer model of the ARM became the formal definition of the processor, and all paper specifications, logic diagrams, transistor layouts and test patterns were validated against it.

The first ARM model was written in BBC BASIC. Global variables were used to represent signal levels or bus values, and procedures performed the functions of the logic blocks on those signals. Typically, each block was represented by two procedures, one for each clock phase. Both clock phases were represented by appropriate sequences of procedure calls. External memory was modelled very simply, using an array initialized with the required program. This model contained no propagation delay information, and could not therefore be used to locate critical timing paths; these had to be checked by hand (assuming that they could first be identified!).

For example, this is the model of the ARM address register:

```
2950 DEF PROCaregph1
2960 tareg%=areg% AND 3
2970 ENDPROC
2980
2990 DEF PROCaregph2
3000 oareg%=tareg%
3010 achkx%=FALSE
3020 IF aregs%=0 areg%=aregn%*4
3030 IF aregs%=1 areg%=alu% AND &3FFFFFF:
                 achkx%=alu% AND &FC000000
3040 IF aregs%=2 areg%=inc% OR areg% AND 3
3050 IF aregs%=3 areg%=reg%(15)
3060 ENDPROC
```

In this model, PROCaregph1 is called for every phase 1 clock, and simply copies the bottom two bits of the address register (areg%) into a temporary transparent latch (tareg%). PROCaregph2 is called for every phase 2 clock; it copies the two bits preserved in phase 1 into a second latch (oareg%), which sends the old value of areg% for use in sorting out the appropriate byte in a byte load. The rest of the model is a multiplexer, which selects the new value of areg% from an exception value (aregn%), a value calculated in the ALU (alu%), the previous value incremented by one word (inc%, with the bottom two bits of areg% unchanged), or the PC (reg%(15)). A boolean (achkx%, address check exception) is used to flag when the ALU result is used and the top six bits are not all zero. The multiplexer is controlled by the two address register select bits (aregs%).

Though this approach was adequate for the first ARM chip (and prc luced working silicon), it was not adequate for the support chips, where timing interdependencies were much more critical. An event driven simulator was constructed (again in BASIC), into which the ARM model was grafted by adding timing information at the periphery. This simulation enabled the first set of four chips to be designed and debugged, though it needed an ARM based system to deliver usable performance when the entire chip set was included in the model.

The importance of the system design tools was recognized during this first phase of development, and a more powerful system level simulator was written in Modula-2 (and later extended in C). The 2 micron ARM2 was modelled using this system, with all its internal timings accurately represented. This allowed a much tighter design to be produced, with a far higher probability of producing a logically correct chip.

This is the Modula-2 model of the ARM address register:

```
TYPE

    ModelOfAreg = POINTER TO RECORD          (* inputs  *)
                        model: ModelCore;    (*   and   *)
                   ph1, ph2, ale: pSignal;   (* outputs *)
           inc, alu, r15, ARegs: pBundle;
    ARegn, Naddr, AReg, oAReg: pBundle;
                        achkx: pSignal;
                END;

    AregState   = POINTER TO RECORD          (*   state  *)
            Nad, NadVal, tAReg: CARDINAL;
                END;
```

```
PROCEDURE AddressReg(modelInst:ADDRESS);
VAR       me: ModelOfAreg;
       state: AregState;
      f, t, a: CARDINAL;
BEGIN
  me := ModelOfAreg(modelInst);
  WITH me^ DO
    state := model.state;
    (*————————————————*)
    IF ph1^.value = One THEN
      state^.tAReg := CARDINAL((AllOnes -
                        BitSet(state^.Nad)) * BitSet(3));
    END; (* if *)
    (*————————————————*)
    a := Zero;  (* By Default *)
    IF ph2^.value = One THEN
      SimQ.BundleEvt(oAReg, Defined, state^.tAReg,
                                  me, 10*NanoSecond);
      SimKernel1.Gather(ARegs);
      CASE ARegs^.binValue OF
        0: SimKernel1.Gather(ARegn);
           t := 4 * ARegn^.binValue; f:= ARegn^.value;
      | 1: SimKernel1.Gather(alu);
           t := alu^.binValue;        f:= alu^.value;
           IF (BitSet(alu^.binValue)*
               BitSet(0FC000000H)) # BitSet{} THEN
             a := One;              (* address exception *)
           END; (* if *)
      | 2: SimKernel1.Gather(inc);
           t := 4 * inc^.binValue + state^.tAReg;
                                  f:= inc^.value;
      | 3: SimKernel1.Gather(r15);
           t := 4 * r15^.binValue;    f:= r15^.value;
      END; (* case *)
      state^.Nad := CARDINAL(AllOnes - BitSet(t));
      state^.NadVal := f;
      SimQ.BundleEvt(AReg, f, t DIV 4,
                                  me, 10*NanoSecond);
    END; (* if *)
    SimQ.SignalEvt(achkx, a, me, 6*NanoSecond);
    (*————————————————*)
```

```
   IF ale^.value = One THEN
      SimQ.BundleEvt(Naddr, state^.NadVal, state^.Nad,
                                           me, 7*NanoSecond);
   END; (* if *)
   (*———————————————————*)
 END; (* with *)
END AddressReg;
```

This model is functionally very similar to the above BASIC model, though here an output latch on the address lines is also modelled. There are many additional statements which manage the simulator environment, and Modula-2 is rather more verbose than BASIC. This model has full timing information, and propagates unknowns from the input buses to the output bus correctly.

All of these tools were design aids at the system level, and were totally disconnected from the transistor level layout and simulation tools. The latter were obtained from VLSI Technology Inc, of San Jose, California, who also fabricated the devices. Higher performance instruction set emulators were also written for software development, so the ARM was modelled in many different ways.

The sequence of activities involved in producing the ARM layout tape is shown in figure 103, and some details are worth highlighting:

(1) The chip was modelled (in BASIC for the 3 micron part, Modula-2 for the 2 micron part) and tested/corrected by the designer until all instructions functioned correctly.

(2) Using higher level instruction set emulators, a validation suite was written by programmers (independently of the design of the system model). The validation suite was extensive, taking a 3 micron ARM two CPU weeks to run it on the Modula-2 model of the 2 micron ARM (and representing about four million modelled CPU clock cycles). This was an independent quality check on the design-correctness of the system model.

(3) Specifications of the various blocks were written and released to the VLSI engineers. They did gate level design, layout, and simulation. Timing figures generated by SPICE simulations of circuits extracted from the physical layout were fed back into the system model to improve its accuracy.

(4) The system model was used to generate block level test pattern files, which were used to validate the logic of the transistor level design. These were reviewed for coverage, untoggled nodes, etc. This

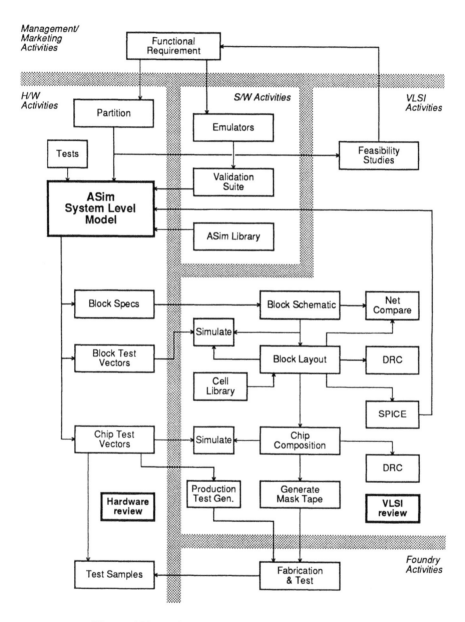

Figure 103: VLSI design methodology within Acorn

procedure ensured that the layout and system block models corresponded to each other functionally.

(5) The system model was used to generate top level chip production test patterns. The complete transistor layout hierarchy was flattened and switch-level simulated against these test patterns. (Note that the closing of the verification loop at block level ensured that no detail design problems were found at chip level simulation, where the diagnosis and simulation cycle times are much longer than they are at block level.)

(6) The system design model, block specifications, and tests were subject to peer group review by the other system designers. The layout and transistor simulation data were likewise reviewed by all the VLSI design group. On all chips some improvements were recommended which increased confidence in the ultimate functionality of the designs.

(7) When all the reviewers were happy, the mask tape was generated and shipped for fabrication. Printed circuit boards were similarly designed, simulated, reviewed, and fabricated. In all cases to date (including both versions of ARM, and the support chips), chips taken from the first untested batch of engineering samples have plugged in and worked.

The above description covers the steps used in the design process, though perhaps it oversimplifies the sequence. For instance, in several cases detail design studies were done at the layout level to establish the feasibility of various design approaches before the system level designer made a final choice. The principle feature of this methodology is the adoption of one model (the system model) which is first validated as fully as is practical by means independent of its designer, and it is then used as a reference against which all other representations are checked.

The transistor level implementation was developed from written specifications of the component blocks of the chip, and we shall now look at some details of the implementation. Firstly though, in order to complete the description of the design methodology, the full block specification of the ARM address register is presented on the next three pages.

Specification for ARM memory address register. (AREG)

30/9/86

Overview:

This block holds the address which is presented to the external memory system. The address may be loaded from the ALU, the PC, or directly from the address incrementer (INC). It may also be forced to a vector address for exception handling. Included in this block is logic which checks the top 6 ALU bits to trap address exceptions.

Signal Definitions.

Buses.

INC[25:2]   Inputs, from INC.

This is the output of the incrementer block, and comes valid during phase 2.

ALU[31:0]   Inputs, from ALU.

The ALU output bus is a potential source of the address, and also comes valid during phase 2.

R15[25:2]   Inputs, from REGBANK.

The PC may be loaded directly into the address register, and when so used will be valid during the whole cycle. (Cycles where the PC is modified will load AREG from the source of the PC, rather than directly from the PC.)

NADDR[25:0]   Outputs, to pins.

The principle outputs from this block are these memory address lines, which go to pins via tri state buffers controlled by ABE.

AREG[25:2]   Outputs, to INC.

These lines represent the internal value of the address register, at the inputs to the latches controlled by ALE. They are the inputs to the INC block, and should be valid throughout phase 1.

Control lines.

AREGS[1,0]   Inputs, from SKP.

These lines determine the source of the address for the next cycle. They are valid during phase 2, and are generated by the SKP block.

ALE   Input, from a pin.

This input comes from off chip, and controls a transparent latch on the NADDR[25:0] outputs to allow the timing to be optimized for dynamic or static memory device

AREGN[2:0]   Inputs, from INT.

These bits are loaded into AREG[4:2] and the rest of AREG[25:5,1:0] is set to zero for exception handling (ie when AREGS[1,0]=0).

OAREG[1,0]   Outputs, to DCTLBW and SCTL.

These outputs reflect an old value of AREG[1,0] and are used to direct data through DIN and SHIFT during the cycle following the memory load. These lines should become valid early in phase 2 and remain so throughout phase 1.

ACHKX   Output, to INT.

This output flags a load from the ALU with a 1 somewhere in ALU[31:26]. This is used to cause an address exception, thereby allowing a primitive implementation of dynamic data type checking.

Clocks.

PH1, PH2

Non overlapping clocks.

Functions.

The function of the block is described with reference to the main internal register, AREG[25:0].

During phase 1 the bottom two bits (AREG[1,0]) are latched into TAREG[1,0].

During phase 2 TAREG[1,0] are latched into OAREG[1,0]

and AREG[25:0] is loaded from the selected source:

```
if AREGS[1,0]=0 then
                AREG[25:5]=0
                AREG[4:2] =AREGN[2:0]
                AREG[1,0] =0
                ACHKX     =0
if AREGS[1,0]=1 then
                AREG[25:0]=ALU[25:0]
                if ALU[31:26]=0 then
                                        ACHKX=0
                                else
                                        ACHKX=1
if AREGS[1,0]=2 then
                AREG[25:2]=INC[25:2]
                AREG[1,0] =OAREG[1,0]
                ACHKX     =0
if AREGS[1,0]=3 then
                AREG[25:2]=R15[25:2]
                AREG[1,0] =0
                ACHKX     =0
```

The addresses seen at the pins are also controlled by ALE, the address latch
enable.

```
if ALE=1 then
        NADDR[25:0]=NOT AREG[25:0]
    else
        NADDR[25:0]=previous value
```

*(ALE should never be low for more than one clock phase, so dynamic latches
may be used.)*

4.6 DATAPATH IMPLEMENTATION

The performance of the CPU should ultimately be limited by the datapath cycle time, so a good way to start the design is to look at the principle components which determine that time. They should be the register bank, the barrel shifter (in the case of ARM, where the barrel shifter is used in series with the ALU in every cycle), and the ALU.

The *register bank* uses three buses; two are used to read operands and the third is used to write the ALU result into a register. The basic register cell is shown in figure 104. The data value is stored on the cross coupled inverter pair. The A and B buses are dynamic, and must be precharged to a high level for reading. Standard logic levels are used (ie no special sense amplifiers), which is practical because of the small number of registers. The stored data value is passed onto a read bus by selectively discharging the bus. Because of the relatively high capacitance of the register read buses, just driving these to the appropriate level takes 25% of the datapath cycle time.

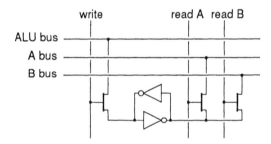

Figure 104: ARM register cell structure

A write operation is performed by overdriving the storage node through the pass transistor. Careful design and simulation are necessary to ensure that this write mechanism will always work so long as the semiconductor processing and operating temperature and voltage are within tolerance. The write operation is very fast and does not contribute significantly to the datapath cycle time.

Each register requires three decoders, one for each bus. These decoders activate the appropriate register according to a field in the current instruction and the currently selected bank (user, supervisor, interrupt or fast interrupt). The main implementation problem is to match the decoders to the spacing of the register cells; the solution on ARM is to use a tall thin decoder layout, and to arrange three of these on top of each other and above the corresponding row of register cells. The register bank floorplan is shown in figure 105.

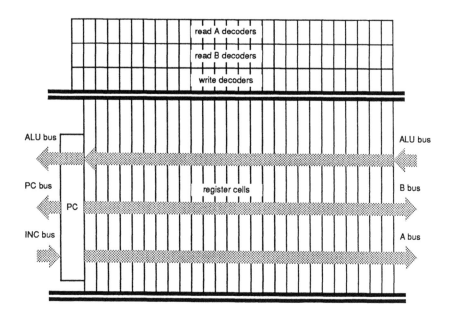

Figure 105: ARM register bank floorplan

The Program Counter is in the register bank, as register 15. It has five ports, unlike the other registers which have only three. This is possible without compromising the size of the standard register cell because the PC can be placed at one end of the bank, and the main read buses do not need to pass out of that end. Therefore those read buses can connect to one end of the PC cell and stop, and the two additional buses can then exploit the space which they would have used to pass through.

The *barrel shifter* operates on one register operand, or a data field in the current instruction, or an incoming data value. It is in series with the ALU on every datapath cycle, and must therefore add very little delay to the operand path to the ALU. It is implemented as a cross switch matrix, where every input can pass to any output through just one pass transistor. Figure 106 illustrates this idea for a 4x4 matrix; ARM uses a similar 32x32 structure. The control decode setup time to enable the correct path is relatively long, but the operand path is very fast, and is responsible for about 15% of the datapath cycle time.

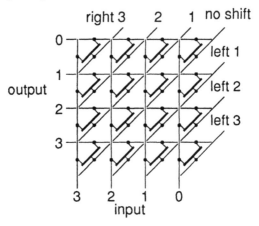

Figure 106: A 4x4 barrel shifter matrix structure

Because the pass transistors are fast when transmitting a low level, but slow to transmit a high level, the outputs of the shifter are treated as dynamic nodes and precharged. It is very important therefore that the control lines are set up before input zeroes start arriving from a register read, otherwise outputs will be discharged erroneously. The decoder to set up the control lines is very complex, however, so it is not possible to guarantee that the correct control line is active before the register read is enabled. (It will, of course, be active before the worst case read time, otherwise the control logic would affect the worst case cycle time, which is not RISC philosophy!)

The solution is to use a clock to turn off all the control lines, and then ensure that only the correct one is enabled. A fast register read will get held up until the control line is on, but a worst case read will find the appropriate route already enabled when it gets to the shifter. The worst case cycle time is never impacted by the control logic, and (more importantly) an output is never erroneously discharged by unexpectedly early data.

The *ALU* performs standard arithmetic and logic operations on the two input operands. All operations are on the full 32 bits. The ALU is constructed from fully static CMOS logic, and the logic for one bit of the ALU is shown in figure 107.

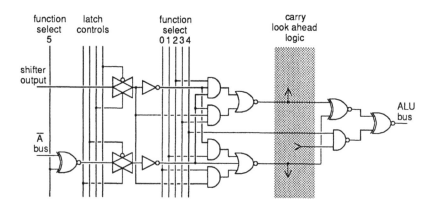

Figure 107: ARM ALU logic for one bit

Note that the arithmetic operation is always an add. Subtraction is performed by logical inversion of the input to be subtracted, and adjusting the input carry accordingly. The various functions of the ALU, and the values of the control lines which generate those functions, are detailed in table 34.

The critical path through the ALU includes a carry chain which is only used for the arithmetic operations. In principle the two least significant bits must be added and a carry out generated (or the carry in propagated) before the next two bits can be added to generate or propagate a carry for the next bit pair, and so on. The first ARM implementation used a simple serial system, carefully designed to have only one gate delay at each stage. The 2 micron ARM uses a carry lookahead scheme (figure 108) to look at groups of four bits to decide if that group will generate a carry out independently of the carry in, or propagate a carry in if there is one. The carry signal can then move across the four bits in one gate delay, reducing the number of gates in the carry path from 32 to 8. More efficient carry paths are possible, but there is always the trade off of performance against design regularity and time, and silicon area. The ARM2 carry chain still represents 15% of the datapath cycle time even with the carry look ahead logic, though the rest of the ALU logic is longer still at 30% of the cycle time.

Table 34. ARM ALU function codes

function select						ALU [31:0]
5	4	3	2	1	0	
0	1	0	1	0	0	A and B
0	1	1	0	0	0	A and not B
0	1	1	0	0	1	A exor B
0	0	1	0	0	1	A plus not B plus carry
0	0	0	1	1	0	A plus B plus carry
1	0	0	1	1	0	not A plus B plus carry
0	1	0	0	0	0	A
0	1	0	0	0	1	A or B
0	1	0	1	0	1	B
0	1	1	0	1	0	not B
0	1	1	1	0	0	Zero

We have now covered those elements of the datapath which form the critical path which determines the cycle time. The cycle time should be determined solely by register read, shift, ALU, and register write times. All other datapath (and control) components should be designed so that they do not affect this critical path. For instance, other sources of data onto the register read buses must be a least as fast as the register bank. This should not present a problem, since the pressure on transistor size is far greater in the register bank, where 1,712 bus driving transistors must be accommodated, than in other blocks where at most 32 are required.

The *field extraction unit* can connect various fields from either the current instruction, or from an incoming data word, to the B bus. In a branch instruction it will extract a 24-bit offset to be added to the PC, in a data operation with immediate operand it will extract the lowest byte from the instruction, and in a byte read it will extract the appropriate byte from the data word. A load or store with base plus offset addressing uses a 12-bit immediate offset, which must be extracted from the instruction. In all cases (except where the full 32 bits are required) the value is zero-extended to 32

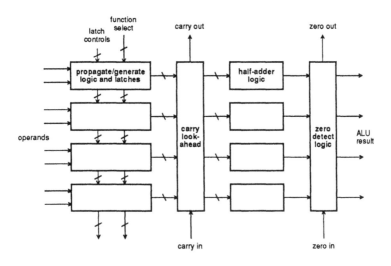

Figure 108: ARM ALU carry lookahead structure

bits, which just amounts to ensuring that the unused bits of the bus are not discharged (since the register read buses both use inverse logic levels).

Note that wherever a negative constant might be expected in an instruction, for instance a negative offset from a register base in a load address calculation, the instruction contains an option to subtract the constant instead of adding it. Negative constants are therefore never required, and sign extension logic is unnecessary. The general structure and functions of the field extraction unit are shown in figure 109. All required fields are accessible by having independent enables on each byte of the word, plus separate enables for the two halves of the second byte (for the 12-bit Load/Store offset field).

The *instruction pipeline* holds the current instruction and instructions already fetched which are awaiting execution. The instruction prefetch system on ARM is intended to allow the processor to be kept busy on continuous data processing instructions, and this requires a three-instruction-deep pipeline to support the fetch/decode/execute stages. When a multi-cycle instruction is encountered the pipeline must freeze temporarily, and allow the next instruction through to the decode stage during the last cycle of the multi-cycle instruction.

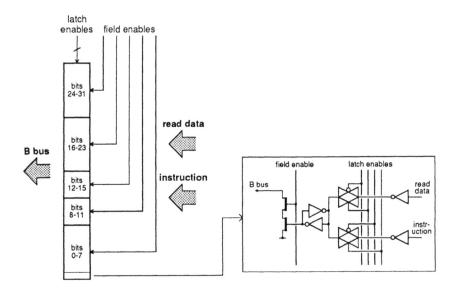

Figure 109: ARM field extraction unit

Figure 110 shows the structure in block diagram form. The various pipeline registers are constructed from transparent latches, with each register having a separately controlled clock. The first register (latch A) is active whenever an instruction fetch occurs, so that all instructions are copied into it. The second (latch B) is activated on the last cycle of an instruction, to allow latch A to be used in the first cycle of the following instruction (which is always an instruction fetch). A multiplexer either selects a new instruction or recycles the old one, and passes it through latch C for decoding during the next cycle. Latch D supplies the current instruction to secondary decoders, and returns it to the multiplexer for multi-cycle instructions. The multiplexer and return latch are used here to perform a function that might more normally be done by conditionally disabling the latch C clock. The conditioning signal is, however, valid too late for the latter approach.

Note that latches A and B are both open at once during a single cycle instruction, so that the fetched instruction is passed directly through for decoding in the next cycle. Normally, therefore, instructions pass through the pipeline block very quickly. The registers are only fully used when a multi-cycle instruction causes the subsequent instructions to wait.

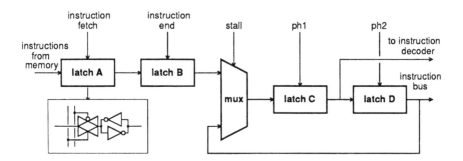

Figure 110: ARM instruction pipeline

The *data output* block is used when writing data to external memory, and it takes the value off the B bus and latches it until the end of the cycle. When a byte value is to be written, the byte is replicated four times across the 32-bit data bus (figure 111). This means that an external byte organized memory can be wired directly to the databus, and the byte can be written to the correct memory block by enabling only that block.

At the other end of the datapath is the *address register* and associated logic. This is used to select the address for the next cycle from the appropriate source (which may be the PC, the ALU, the address incrementer, or an exception address) and to latch it appropriately. The block is constructed as a four input multiplexer (figure 112). To allow flexibility in the address timing an externally controlled latch is included. The address incrementer is the most frequently used address source, and it simply generates a new address value which is one word on from the previous address. Whenever it is used to generate sequential instruction addresses, the incremented value is copied back into the PC every cycle to ensure that the register 15 copy of the PC is up to date.

The final piece of logic on the datapath is a *two-bit multiplier shifter*, which is used to support the multiply instructions. It presents the next two bits of one operand to a section of control logic which performs a two bits at a time Booth's multiplication algorithm. The shifter also detects when all the ones in the operand have been used, and this causes the termination of the multiply. The implementation of this block is based on simple transparent latches.

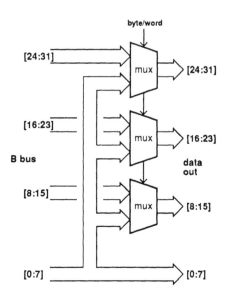

Figure 111: ARM data output block

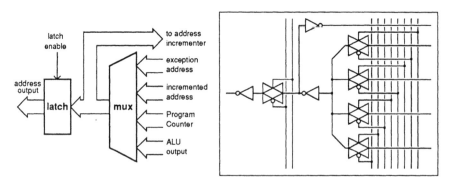

Figure 112: ARM address register

4.7 CONTROL LOGIC IMPLEMENTATION

The general layout of the 2 micron ARM chip is shown in figure 113. This shows that the datapath of the processor occupies just under half of the active chip area. The remainder of the chip contains the logic which controls the datapath, and causes it to perform the operations required by each instruction.

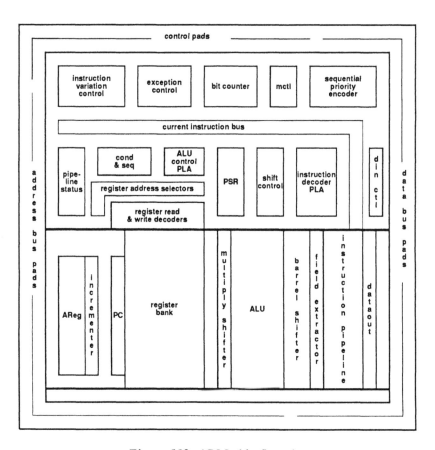

Figure 113: ARM chip floorplan

The datapath of a processor is where the work gets done. It is important that the datapath cycle time is minimized, and that the control logic does not interfere with this minimization. There is no point, however, in making the control logic any faster than is necessary to keep the datapath cycling at its maximum speed. Therefore whereas the datapath is designed to be as fast as possible, the control logic need only be as fast as the datapath. This allows the design methodology for the control logic to be more orientated towards fast implementation than towards fast operation. In general, the control logic on ARM was built using elements from a library of standard cells, whereas the datapath was built from fully customized cells.

An exception to the 'as fast as the datapath' rule arises when the control logic in question is determining the timing of an external signal; this might be either the setup requirements of an input, or the delay time to when an output becomes valid. In such cases, rather than making assumptions about the capabilities of the external control logic, the 'as fast as possible' approach is retained.

The datapath elements have been designed to support the data operations and routing required by the instruction set, as described in the previous section. All that is needed is to take the current instruction at the top of the pipeline and decode it to set up the appropriate datapath actions in each cycle.

A *horizontal* decode structure would attempt to take each instruction and decode it in a PLA or ROM to generate the datapath control signals directly. It is clearly not practical, however, to fully decode 2^32 possible instructions individually. ARM uses a two level decode structure:

(1) The instruction decoder PLA looks at a few key bits from latch C in the pipeline. These bits (in fact bits 4, 7, 20, and 24 to 27) are sufficient to differentiate the instruction set into classes, where all members of a particular class have the same characteristics in terms of bus usage, number of cycles, and so on. The instruction decoder outputs information along the lines of 'in this cycle the B bus source is specified in bits 16 to 19 of the current instruction', and 'the shifter should always be shifting left two bits here'.

(2) The secondary decoding is distributed close to the datapath blocks, and uses the above information to control a multiplexer (for instance) which selects bits 16 to 19 of the instruction and sends them to the B bus read decoder in the register bank. The secondary decoder has hardware to cope with all the possible operations the instruction set expects it to do. It gets broad hints from the instruction decoder about which function to choose, and it fills in the details itself by directly accessing the bits of the current instruction that it needs to know.

By way of example, consider the control logic for the field extraction block (which is labelled 'din ctl' in the floorplan). The operations which are required of this block are:

(1) Do not drive the B bus at all.

(2) Extract the bottom byte from the current instruction and place it onto the B bus. This is used by data operations with immediate operands, and by coprocessor data transfers.

(3) Extract the bottom 12 bits from the current instruction and place them onto the B bus. This is used by single register transfers.

(4) Extract the bottom 24 bits from the current instruction and place them onto the B bus. This is used by branch instructions.

(5) Place the entire loaded data word onto the B bus. This is used by single and multiple word load instructions, and coprocessor register transfers to ARM.

(6) Extract the addressed byte from a loaded data word and place onto the B bus. This is used by byte load instructions.

Note that most of these are fixed operations, but the last one requires information from the preceding address calculation. The implementation (figure 114) is based on a four input pass-transistor multiplexer which selects between the 8-bit, 12-bit, 24-bit or load options. The multiplexer select controls come from the central instruction decoder. The 8-bit, 12-bit and 24-bit options are hardwired into the multiplexer, and the load option selects information from a separate decoder which uses the bottom address bits and the byte/word flag from the previous cycle (when the memory transfer of the data item took place).

The 'do nothing' option is selected by an override on the multiplexer outputs. This detects when either the register bank or the program status register is driving the B bus, and it disables the field extractor in either case. The field extractor therefore always drives the B bus when the bus is not explicitly required for other purposes.

The phase 1 clock is gated into the final stage of the decode to ensure that the field extractor only drives the bus during phase 1. The bus is pre-charged high during phase 2 by logic in the register bank.

The control blocks for the register addresses are similar simple multiplexers. The control block for the shifter is similar in principle, though more complex in detail, since it contains an 8-bit latch for register controlled shifts and a

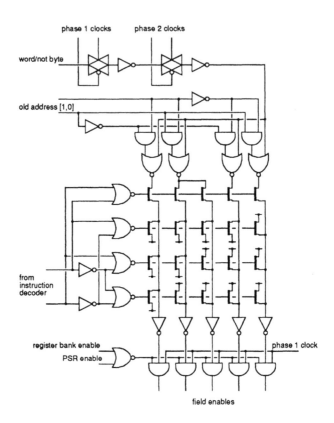

Figure 114: The control logic for the ARM field extraction unit

fair amount of special case logic to control the shift carry out.

The ALU control block is a PLA. This decodes explicit ALU operations (from data operation instructions) and implicit ones (for example adding a branch offset) into the ALU function selects described earlier. It also has an output which indicates whether a particular operation produces a result (to differentiate CMP from SUB, for example), and another which indicates whether an operation is arithmetic or logical. The multiply Booth's algorithm state machine is encoded partly into this PLA, and partly in dedicated logic ('mctl' in the floorplan).

The condition code evaluation block ('cond') is a multiplexer which decodes the top four bits of the instruction to select the appropriate logical combination of the PSR flags. The PSR flags become valid later than the instruction bits, so the logic was designed to decode the latter in time to propagate the former quickly. This block produces a single output which indicates condition pass or fail; the fail condition terminates the current instruction.

The sequence controller ('seq') gives a cycle number to the instruction decoder during multi-cycle instructions, and goes into a fixed state for the open ended instructions, awaiting a termination signal from the priority encoder (for multiple register transfers), the multiply state machine (for multiplies), or a coprocessor (for the coprocessor busy-wait loop and coprocessor data transfers).

The pipeline status register contains two bits of information about each instruction in the pipeline:

(1) *A valid bit.* This indicates whether the PC has been directly modified since the instruction was fetched. If it has, the instruction should be ignored.

(2) *An abort bit.* If the memory controller signals a memory fault when an instruction is fetched, the abort bit is set. If the instruction enters the decode phase, it will not be executed. Instead, the prefetch abort sequence will be entered. Note that if the instruction is not executed, the abort sequence will not be entered, so if a branch is at the end of a memory page, the prefetching of the two instructions which follow the branch cannot cause an abort so long as the branch is taken.

The pipeline status register operates very much like the instruction pipeline itself. The abort bit is sampled during the fetch cycle along with the associated instruction, and the valid bit is set when the instruction is fetched, and then cleared throughout the pipeline if the PC is modified.

The priority encoder and bit counting logic are used only by the load and store multiple register instructions, where they process the information in the bottom sixteen bits which specify the registers to be transferred.

The Program Status Register (PSR) contains the ALU flags, the interrupt disable flags, and the processor mode bits. This is where the protected supervisor mode is implemented. The block is implemented as two 8-bit latches which pass the information from one to the other on one clock phase, and back to the first latch on the next clock phase. The contents are modified when appropriate by multiplexers on the latch inputs, and the mode interlocks are implemented by restricting the multiplexer options in user mode.

The exception control block prioritizes any pending exceptions, and takes over the instruction decoder at the end of the current instruction. It sends the exception vector address to the address register, and the mode required for handling the exception to the PSR. Asynchronous exceptions (the interrupts and reset) are synchronized to the processor clocks before they enter the prioritization logic.

The last control block handles variations in instructions. For example, during a data operation the address for the next cycle is taken from the address incrementer. This is correct unless the data operation modifies the program counter, in which case the address should take the modified value, preferably directly from the ALU. The address register select lines therefore come from this block. The central instruction decoder sends information which specifies the address source as the incrementer if the PC is not being modified, the ALU if it is. The instruction variation block makes the final choice. Likewise the register bank write enable is generated here, so that a load register instruction will write to the destination unless there data transfer caused an abort. There are many similar special cases.

Most of the control logic is implemented using gates from a standard cell library. There is very little regularity to be exploited in the layout, so this approach is a practical solution to the problem of laying out large numbers of gates of random logic in a reasonable time. The standard cell based designs are also relatively easy to modify, which is useful in the later stages of design as the last few design bugs are squeezed out. The most obscure problems tend to be in infrequently used areas of the control logic, so the ease of correction of the control logic is an important aspect of the design methodology.

4.8 TIMING CHARACTERISTICS

The ARM execution unit is not pipelined; a single cycle includes register read, operand shift, ALU operation and result write. The fetch and decode stages are pipelined, however, so that while the execution unit is performing one instruction its successor is being decoded and the one after that is being fetched from memory. ARM can therefore complete an instruction every cycle (see figure 115). Load and store instructions take additional cycles for the data transfers, and branches take three cycles because they break the pipeline operation. Most RISC processors use delayed branching to avoid the pipeline break, but the complexity of restarting after an abort in a delayed instruction was felt to be too great. Instead, all ARM instructions may be skipped conditionally, so that short forward branches can be omitted altogether.

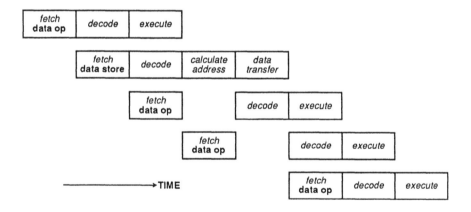

Figure 115: An example of ARM pipelining

The timing of the internal and external signals is shown in figure 116. This chart was constructed by following all the paths which contribute to each signal, and identifying the critical paths by using propagation times generated by SPICE analogue simulations of each component of each path. The period of validity of each signal is indicated by the solid horizontal line, and the critical paths are the arrows which connect the start points of certain lines together. External inputs are set up to meet the latest possible times which satisfy all the internal requirements.

It was much more straightforward to produce this timing chart for the ARM CPU than it was to produce equivalent documentation for the ARM support chips. The simple clocking and pipeline schemes implemented on the CPU contribute greatly to the ease of understanding of the chip.

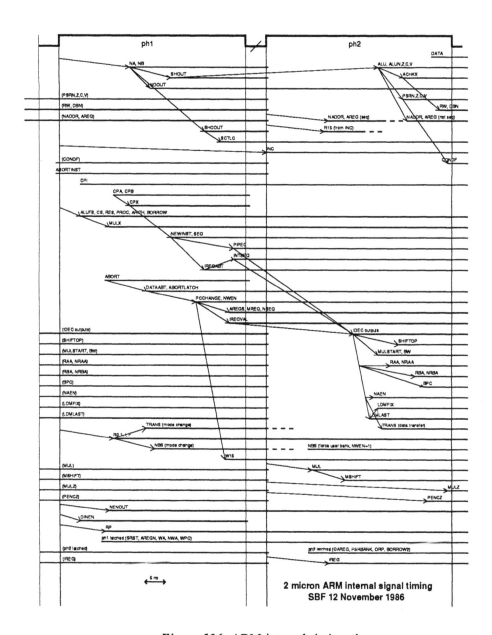

Figure 116: ARM internal timing chart

4.9 THE MEMORY INTERFACE

ARM reads instructions and data from, and writes data to, its main memory via a 32-bit data bus. A separate 26-bit address bus specifies the memory location to be used for the transfer, and the $\overline{\text{R}}$/W signal gives the direction of transfer (ARM to memory or memory to ARM). Control signals give additional information about the transfer cycle, and in particular they facilitate the use of DRAM page mode where applicable. (Interfaces to static RAM based memories are not ruled out; they are in general much simpler than the DRAM interface described here.)

4.9.1 Cycle Types

All memory transfer cycles can be placed in one of four categories:

(1)   *Non-sequential cycle.* ARM requests a transfer to or from an address which is unrelated to the address used in the preceding cycle.

(2)   *Sequential cycle.* ARM requests a transfer to or from an address which is either the same as the address in the preceding cycle, or is one word after the preceding address.

(3)   *Internal cycle.* ARM does not require a transfer, as it is performing an internal function and no useful prefetching can be performed at the same time.

(4)   *Coprocessor register transfer.* ARM wishes to use the data bus to communicate with a coprocessor, but does not require any action by the memory system.

These four classes are distinguishable to the memory system by inspection of the **MREQ** and **SEQ** control lines (see table 35). These control lines are generated during phase 1 of the cycle before the cycle whose characteristics they forecast, and this pipelining of the control information gives the memory system sufficient time to decide whether or not it can use a page mode access.

Figure 117 shows the pipelining of the control signals, and suggests how the DRAM address strobes ($\overline{\text{RAS}}$ and $\overline{\text{CAS}}$) might be timed to use page mode for S-cycles. Note that the N-cycle is longer than the other cycles. This is to allow for the DRAM precharge and row access time, and is not an ARM requirement. When an S-cycle follows an N-cycle, the address will always be one word greater than the address used in the N-cycle. This address (marked "a" in the above diagram) should be checked to ensure that it is not the last in the DRAM page before the memory system commits to the S-cycle. If it is

Table 35. Memory cycle types

MREQ	SEQ	Cycle type	
0	0	Non-sequential cycle	(N-cycle)
0	1	Sequential cycle	(S-cycle)
1	0	Internal cycle	(I-cycle)
1	1	Coprocessor register transfer	(C-cycle)

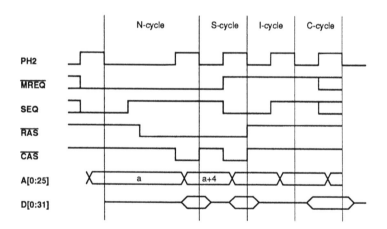

Figure 117: The timing characteristics of each cycle type

at the page end, the S-cycle cannot be performed in page mode and the memory system will have to perform a full access. The processor clock must be stretched to match the full access.

When an S-cycle follows an I- or C-cycle, the address will be the same as that used in the I- or C-cycle. This fact may be used to start the DRAM access during the preceding cycle, which enables the S-cycle to run at page mode speed whilst performing a full DRAM access (figure 118).

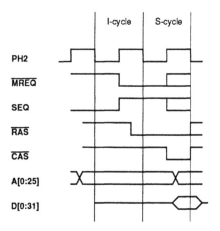

Figure 118: Merged I- (or C-) and S-cycles

4.10 THE COPROCESSOR INTERFACE

The functionality of the ARM instruction set may be extended by the addition of up to 16 external coprocessors. When the coprocessor is not present, instructions intended for it will trap, and suitable software may be installed to emulate its functions. Adding the coprocessor will then increase the system performance in a software-compatible way.

Three dedicated signals control the coprocessor interface, $\overline{\text{CPI}}$, CPA and CPB. An external pull-up resistor is normally required on both CPA and CPB to support multiple off-chip coprocessors.

Coprocessor Present/Absent

ARM takes $\overline{\text{CPI}}$ low whenever it starts to execute a coprocessor (or undefined) instruction. (This will not happen if the instruction fails to be executed because of the condition codes.) Each coprocessor will have a copy of the instruction, and can inspect the CP# field to see which coprocessor it is for. Every coprocessor in a system must have a unique number, and if that number matches the contents of the CP# field, the coprocessor may pull the CPA (coprocessor absent) line low. (It may also decline the instruction by not doing so, so a coprocessor may implement only part of its instruction space.) If no coprocessor has a number which matches the CP# field, CPA will float high, and ARM will take the undefined instruction trap. Otherwise ARM observes the CPA line going low, and waits until the coprocessor is not busy.

Busy-Waiting

If CPA goes low, ARM will watch the CPB (coprocessor busy) line. Only the coprocessor which is pulling CPA low is allowed to drive CPB low, and it should do so when it is ready to complete the instruction. ARM will busy-wait while CPB is high, unless an enabled interrupt occurrs, in which case it will break off from the coprocessor handshake to process the interrupt. Normally ARM will return from processing the interrupt to retry the coprocessor instruction.

When CPB goes low, the instruction continues to completion. This will involve data transfers taking place between the coprocessor and either ARM or memory, except in the case of coprocessor data operations which complete immediately the coprocessor ceases to be busy.

All three interface signals are sampled by both ARM and the coprocessor(s) on the rising edge of PH2. If all three are low, the instruction is committed to execution, and if transfers are involved they will start on the next cycle. If $\overline{\text{CPI}}$ has gone high after being low, and before the instruction is committed, ARM has broken off from the busy-wait state to service an interrupt. The instruction may be restarted later, but other coprocessor instructions may come sooner, and the coprocessor should discard the instruction.

Pipeline Following

In order to respond correctly when a coprocessor instruction arises, each coprocessor must have a copy of the instruction. All ARM instructions are fetched from memory via the main data bus, and coprocessors are connected to this bus, so they can keep copies of all instructions as they go into the ARM pipeline. The $\overline{\text{OPC}}$ signal indicates when an instruction fetch is taking place, and **PH2** gives the timing of the transfer, so these may be used together to load an instruction pipeline within the coprocessor.

Data Transfer Cycles

Once the coprocessor has gone not-busy in a data transfer instruction, it must supply or accept data at the ARM bus rate (defined by **PH2**). It can deduce the direction of transfer by inspection of the L bit in the instruction, but must only drive the bus when permitted to by **DBE** being high. The coprocessor is responsible for determining the number of words to be transferred; ARM will continue to increment the address by one word per transfer until the coprocessor tells it to stop. The termination condition is indicated by the coprocessor releasing **CPA** and **CPB** to float high.

There is no limit in principle to the number of words which one coprocessor data transfer can move, but by convention no coprocessor should allow more than 16 words in one instruction. More than this would worsen the worst-case ARM interrupt latency, as the instruction is not interruptable once the transfers have commenced. At 16 words, this instruction is comparable with a block transfer of 16 registers, and therefore does not affect the worst case latency.

Register Transfer Cycle

The coprocessor register transfer cycle is the one case when ARM requires the data bus without requiring the memory to be active. The memory system is informed that the bus is required by ARM taking both $\overline{\text{MREQ}}$ and **SEQ** high. When the bus is free, **DBE** should be taken high to allow ARM or the coprocessor to drive the bus, and a **PH2** cycle times the transfer.

Privileged Instructions

The coprocessor may restrict certain instructions for use in supervisor mode only. To do this, the coprocessor will have to track either the $\overline{\text{TRANS}}$ pin or the $\overline{\text{M}}[0,1]$ pins.

As an example of the use of this facility, consider the case of a floating-point coprocessor (FPU) in a multi-tasking system. The operating system could save all the floating-point registers on every task switch, but this is inefficient in a typical system where only one or two tasks will use floating-point operations. Instead, there could be a privileged instruction which turns the FPU on or off. When a task switch happens, the operating system can turn the FPU off without saving its registers. If the new task attempts an FPU operation, the FPU will appear to be absent, causing an undefined instruction trap. The operating system will then realize that the new task requires the FPU, so it will re-enable it and save FPU registers. The task can then use the FPU as normal. If, however, the new task never attempts an FPU operation (as will be the case for most tasks), the state saving overhead will have been avoided.

Idempotency

A consequence of the implementation of the coprocessor interface, with the interruptable busy-wait state, is that all instructions may be interrupted at any point up to the time when the coprocessor goes not-busy. If so interrupted, the instruction will normally be restarted from the beginning after the interrupt has been processed. It is therefore essential that any action taken by the coprocessor before it goes not-busy must be idempotent, i.e. must be repeatable with identical results.

For example, consider a FIX operation in a floating-point coprocessor which returns the integer result to an ARM register. The coprocessor must stay busy while it performs the floating-point to fixed-point conversion, as ARM will expect to receive the integer value on the cycle immediately following that where it goes not-busy. The coprocessor must therefore preserve the original floating-point value and not corrupt it during the conversion, because it will be required again if an interrupt arises during the busy period.

The coprocessor data operation class of instruction is not generally subject to idempotency considerations, as the processing activity can take place after the coprocessor goes not-busy. There is no need for ARM to be held up until the result is generated, because the result is confined to stay within the coprocessor.

4.11 CONCLUDING REMARKS

In this chapter we have looked in some detail at the implementation details of the Acorn RISC Machine, and the design methodologies which were used to produce it. It is not claimed that these methodologies are unique or original, or that they are better than other approaches. They are simply the methodologies that evolved in the course of building the ARM and the associated chipset.

The 3 micron ARM was the first fully customized chip designed at Acorn Computers, and was fully operational in April 1985. It was modelled in BASIC, and the system design methodology grew with the design. The fact that first silicon worked and was sold on to end users is due in part to the simplicity of design which the RISC approach allows, but also to the quality of the CAD system supplied by VLSI Technology Inc., and the professionalism of the Acorn VLSI design group led at that time by Robert Heaton.

The 2 micron ARM was the fifth fully customized chip designed at Acorn, and the fifth to work from first silicon. This chip is 5.5mm square, which is very small by the standards of other 32-bit CPUs. Its worst case performance is 12 million data operations per second, and its average performance is 6 MIPS. When it is attached to dynamic memories without any cache, the performance is reduced to around half this level. Its small size is a result of the low transistor count (25,000), and not because of unusually tight layout. It has good real-time performance, virtual memory support, and a protected supervisor mode. This combination of features would not have been possible with the small transistor count without adopting a RISC approach.

Commercial RISC CPUs are generally targetted at maximizing performance. The ARM shows that the RISC approach can also be applied successfully lower down the cost curve, to produce very cost-effective processors for application in personal computers and single board controllers, or as the processor cell in a highly integrated micro-controller unit. The same CPU can also be used with an on-chip cache in very cost-effective personal workstations.

References

Tredennick, N., (1987). Microprocessor Logic Design, the Flowchart Method, Digital Press.

VSLI Technology, Inc., (1987). VL86C010 RISC family data manual, Application Specific Logic Products Division, 8375 South River Parkway, Tempe, AZ 85284.

Wills, J. D., (1988). Personal communication.

5
Further RISC Research

In this chapter we look at some of the RISC research currently taking place which may indicate the way that architectures and organizations will continue to evolve to enable higher performance levels to be achieved.

We look first to academia for possible pointers to the direction of future developments. The examples here are research projects at Berkeley and Stanford universities, the institutions which led the way with the introduction of the RISC philosophy. Both universities are continuing with architectural research, and have higher performance designs in development.

Although the new university architectures have new instruction sets, the emphasis of the work has moved away from the instruction set itself. Both new designs include caches as essential components of the design, and both are targeted at multi-processor configurations. The CPU architecture is important, but the cache organization is recognized as the area where real performance advances will be made possible.

In order to throw light on some of the issues which affect cache organization, a description is presented of the decision process which led to the design of the cache on ARM3, a development of the ARM processor described in chapters 3 and 4. Performance, power dissipation and chip area were all determining factors in the design of this commercial RISC processor with on-chip cache.

Modelling systems have been developed which allow the impact of various architectural decisions to be evaluated, and these systems may guide future architectural directions. Some of the results of one such study are used to reflect on the existing variety of designs discussed in chapters 2 and 3.

Finally, the book is closed with a little speculation on the likely developments of the near future. Single chip RISC processors can be developed on existing processes to offer 100 MIPS performance, low system cost and low power, but possibly not all three at once!

5.1 THE BERKELEY SPUR PROJECT

The SPUR project at Berkeley is directed towards the construction of a multiprocessor system optimized for the execution of programs written in Common Lisp. (SPUR stands for Symbolic Processing Using RISC.) A typical machine has six to twelve processor clusters on a shared NuBus, giving an order of magnitude more processing power than previous workstations.

The SPUR processor cluster contains a RISC CPU which is similar to RISC II, especially in the register organization. It has an on-chip 512 byte instruction buffer (cache). The cluster also has a floating-point chip, a cache controller chip, and the 128 Kbyte cache memory itself.

The CPU itself is unusual in that it handles 40-bit data items, where the top eight bits are an identification tag which is used for dynamic type checking during program execution. This feature is included to improve the efficiency of Lisp execution. The other unusual feature of the cluster is the use of in-cache address translation; the TLB is folded into the main cache.

Cached multiprocessors suffer from difficulties with cache inconsistency. The SPUR architecture handles this problem in hardware, and because the TLB is part of the cache, TLB consistency is ensured by the same mechanism.

5.1.1 Architecture

SPUR retains the register window organization of RISC II (see chapter 2), though the registers are now all 40 bits wide.

The instruction formats are all 32 bits long, with fixed positions for the register specifiers. The instruction set is listed in table 36.

The SPUR instructions represent a move away from the original RISC goal of a simple generic instruction set. The inclusion of instructions which are only useful in symbolic processing has added to the complexity of the processor, and the register bank has increased in size by 25% to accommodate the tags. It will be interesting to see how valuable these features are, and whether they are copied on future commercial CPU designs.

The In-Cache Translation Mechanism

SPUR uses a novel address in-cache translation mechanism (Ritchie, 1985). The memory mapping scheme is a conventional two-level page table, but instead of a translation lookaside buffer for recently used page table entries, these are held in the main cache along with instruction and data items.

Table 36. SPUR instruction set

Arithmetic and logical operations

AND,OR,XOR,ADD,SUB	
INSERT,EXTRACT	*(bytes)*
S	*(shift left logical, right arithmetic or logical)*

Transport operations

LOAD_32	*(load at base + 14-bit signed offset or base + index)*
STORE_32	*(store at base + signed 14-bit offset)*
TEST_AND_SET	*(semaphore primitive)*
LOAD,STORE_EXTERNAL	*(for cache control operations)*

Control transfer

CMP_BRANCH_DELAYED	*(conditionally branch PC relative; optionally nullify next instruction if branch not taken)*
JMP	*(jump to word address, or register + 14-bit signed offset or index)*
CALL	*(call to word address)*
RET	*(conditional return to register + 14-bit signed offset or index)*

Special instructions

READ,WRITE_SPECIAL	*(to access special registers and the kernel PSW)*

(continued)

Table 36. (continued)

Lisp Instructions

LOAD_40	*(load tagged at base + 14-bit signed offset or base + index)*
STORE_40	*(store tagged at base + signed 14-bit offset)*
CAR/CDR	*(as LOAD_40 but with type checking)*
READ,WRITE_TAG	*(transfer a tag to/from the data area of a register)*
TAG_COMPARE_AND_BRANCH	
COMPARE_AND_TRAP	
TAG_COMPARE_AND_TRAP	

Floating-Point Instructions

Loads and stores of single-, double- and extended-precision numbers
Add, subtract, multiply, divide, compare, negate, absolute value
Format conversion

(Source: adapted from Hill et al, 1986)

SPUR produces 32-bit virtual addresses. The top two bits are used to index into one of four 8-bit segment registers, to produce a 38-bit global virtual address. This gives a large enough address space for all processes (on all processors in a multiprocessor system) to share a common global virtual address space. Normally, a memory reference will hit in the cache, allowing an access as in figure 119.

If there is a cache miss, an attempt will be made to locate the page table entry (PTE) in the cache by synthesizing a virtual page table address as shown in figure 120. Note that the virtual page table address uses the full virtual page number. This is therefore a reference to a virtual single level page table. A physical single level page table would be impractical, because it would all have to exist in physical memory, and each process (however small) would require 4Mbytes of page table. A virtual single level page table is quite practical, however, because it does not need to exist in any physical

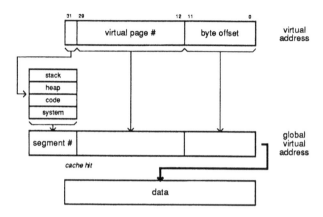

Figure 119: Normal cache access
(Adapted from Ritchie, 1985)

form. In this scheme the virtual single level page table is quite abstract; there is no need for the system to maintain mappings between the physical and virtual page tables.

If the PTE is also absent from the cache, an attempt will be made to locate the address of the page table base in the cache as shown in figure 121. This, again, is a reference to a (different) virtual page table, now only using the top eight bits of the virtual page number. This table is a virtual copy of the physical root page table. If this cache reference succeeds, the PTE can be fetched from physical memory. It will also be copied into the cache so that the next cache miss in this page will find it, and complete without this additional memory reference. The data will then be fetched and cached.

If the page table base address is also absent from the cache, the full physical access sequence (as in figure 122) will be followed, with cache entries created at each stage for the page table base, the page table entry, and the data itself.

The above mechanism can be thought of either as a conventional double-level page table in physical memory, or as a single level table in virtual memory. These two views allow the virtual cache to contain recently used first and second level page table entries, mixed with instruction and data entries. No special flags are needed to identify PTEs; they are present in physical memory as conventional data items identified only by their physical

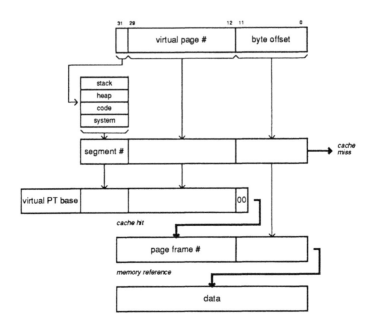

Figure 120: Page table entry found in cache
(Adapted from Ritchie, 1985)

addresses, and they are likewise identified in the cache only by their virtual addresses. Another view is that the virtual page table base is simply a convenient tag to identify the PTE uniquely in the mixed cache.

The in-cache translation mechanism has the beneficial property of automatically balancing the amount of on-chip memory used for page table entries, instructions and data according to current requirements. A separate TLB imposes a fixed partition, which is less flexible. The in-cache approach also allows cache and TLB consistency to be maintained with a single snooping mechanism.

The drawback of this approach is that the translation accesses must be done serially with the cache access, whereas a separate TLB allows translation to proceed in parallel with the cache access. A separate TLB therefore reduces the cache miss time penalty.

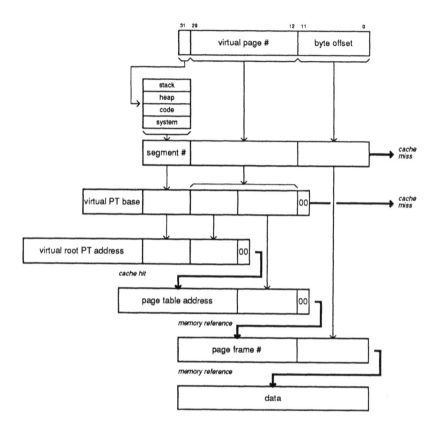

Figure 121: Page table root found in cache
(Adapted from Ritchie, 1985)

5.1.2 Organization

The organization of a SPUR processing node is shown in figure 123. Each node contains three customized components, the CPU, the floating-point unit, and the cache controller. In addition there is the cache itself, which uses several standard static RAMs.

Details of the internal datapath organization of the CPU are given in Lee (1986). The interface to the floating-point coprocessor is described by Hansen and Kong (1986).

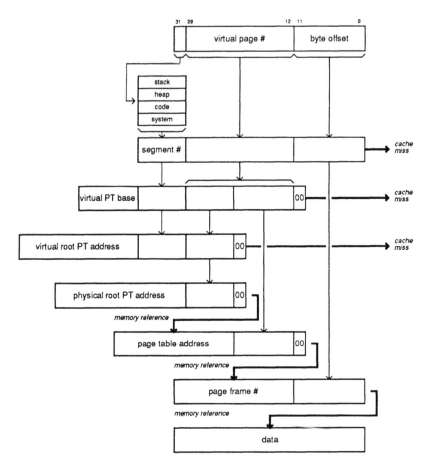

Figure 122: Full access sequence
(Adapted from Ritchie, 1985)

The CPU has an on-chip 512 byte instruction cache which is direct mapped with a line size of eight words. It can reload individual words (each word entry has a valid bit) but a prefetch mechanism continues fetching from the previous loaded instruction address until the external cache is needed for a different instruction stream or a data access.

The off-chip 128 Kbyte cache stores mixed instruction, data, and page table entries. It is direct mapped with a 32-byte line size, and always transfers

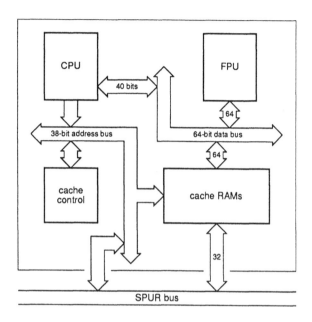

Figure 123: SPUR CPU cluster
(From Patterson, 1987)

aligned 32-byte blocks to and from memory. It is virtually addressed, with a hardware coherency mechanism based on the Berkeley ownership protocol (see chapter 1, or Katz et al, 1985). Synonyms (two different virtual addresses which map to the same physical address) are disallowed in the SPUR architecture definition. Processes which share areas of memory do so by both allocating the same virtual address to the shared area.

5.1.3 Conclusions

The SPUR architecture incorporates a number of important ideas:

- *In-cache address translation.* The folding of the TLB into the cache is an elegant way of reducing the overall complexity of the processing node, and has the additional benefit of giving TLB consistency for free. The drawbacks are small. The time to translate an address is increased, as a result of the translation coming in series rather than in parallel with the cache access. This penalty is reduced as the CPU clock speed increases, and may be lost in bus arbitration time in a multiprocessor system. The other drawback is the reduction in cache hit ratio caused by cache pollution with page table entries. This is reduced as the cache gets larger. Both of these disadvantages are minimal on SPUR, and reduce with future process advances. This feature is therefore likely to find favour with other designers.

- *Cache consistency protocols.* These are developing rapidly, and it is unlikely that the Berkeley protocol will be the last word on the matter. It is important that institutions like Berkeley perform research on such areas, since their record for thorough investigation and publication suggests that we will all become much better informed as a result.

- *The 40-bit register set* and the special instructions for manipulating tags are of use only in systems which make extensive use of tagged data items, which at present restricts their use to artificial intelligence languages such as LISP. Their widespread adoption outside the specialist AI market seems unlikely.

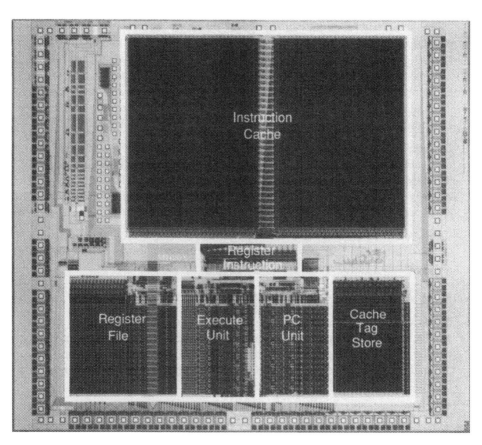

Figure 124: Photomicrograph of MIPS-X
(Copyright ©Stanford University)

5.2 MIPS-X

The MIPS-X processor is a successor to the MIPS project at Stanford University, and is a single VLSI chip containing a 32-bit CPU and 2Kbyte instruction cache. Like the commercial R2000 processor described in chapter 3, MIPS-X calls heavily on the experience gained from MIPS.

MIPS-X is designed to take advantage of the improvements in semiconductor processing which have become available since MIPS. It uses a 2 micron CMOS process, and employs 150,000 devices on an 8mm by 8.5mm die. It was designed to operate at 20MHz with a simple two-phase clock. At this clock rate, the peak bandwidth requirement of the processor is 160 Mbyte/s. This is hard to achieve across a single word wide bus interface, so the on-chip instruction cache is employed to reduce the instruction traffic by 90%.

In addition to the on-chip instruction cache, MIPS-X uses a 256Kbyte off-chip mixed instruction and data cache. The aim is to connect between six and ten processors (each with its own on- and off-chip caches) in a shared memory multiprocessor configuration, giving a machine two orders of magnitude faster than a VAX-11/780.

Compared with the original MIPS design, MIPS-X uses a deeper pipeline, a simpler instruction set, and starts a new instruction every cycle instead of every second cycle. (See Horowitz et al, 1987 for further details.)

5.2.1 Architecture

MIPS-X has 32 general-purpose user registers (increased from the 16 on MIPS), a multiply/divide special register, a chain of PC registers, and a status register. The status register does not contain condition codes; these are only generated transiently during conditional branch instructions. General register 0 always contains the value zero.

The instruction formats are shown in figure 125. The source registers are always specified by the same bits, so that operand fetching can proceed in parallel with instruction decoding. The formats are very simple compared with the original MIPS instruction set; code density was given a much lower priority than decoding simplicity. The instruction set is listed in table 37.

Forwarding logic allows the ALU result of one instruction to be used as an operand of the next. A value loaded from memory is not available to the next instruction, but may be used by the one after that. The single load delay slot has no hardware interlock, so a code reorganizer is used (as on MIPS and R2000) to avoid data race hazards on loads.

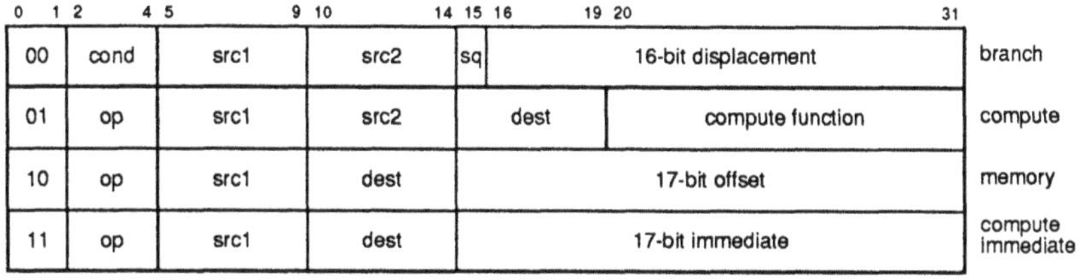

0 1 2 4 5 9 10 14 15 16 19 20 31	

00	cond	src1	src2	sq	16-bit displacement		branch
01	op	src1	src2	dest	compute function	compute
10	op	src1	dest	17-bit offset		memory
11	op	src1	dest	17-bit immediate		compute immediate

Figure 125: MIPS-X instruction formats
(From Horowitz et al, 1987. ©1987 IEEE.)

Branches have two delay slots, with optional nullification if the branch is not taken. The instruction set makes the static assumption that branches are likely to be taken, and therefore the two instructions which are at the branch target can always be moved into the delay slots.

5.2.2 Organization

The internal organization of MIPS-X is shown in figure 126.

The pipeline operates in five stages:

(1) Instruction fetch.

(2) Register fetch.

(3) Execute.

(4) Access memory.

(5) Write result.

Conditional branch instructions use the main ALU to compare the two registers and evaluate the condition whilst calculating the branch target in a separate adder in the PC unit. Either the target or the incremented PC is driven out as the next instruction address at the end of the execute phase, according to the result of the condition. This gives rise to the double branch delay slots.

The on-chip instruction cache is 8-way associative, with 32 lines of 16 words each. The block size for reloading is two words, and each word has an associated valid bit. A cache access takes one cycle; a cache miss causes access to an off-chip mixed instruction and data cache to reload 2 words with a time penalty of two lost cycles.

Table 37. MIPS-X instructions

Compute Instructions

add	*(register or 17-bit immediate)*
sub	*(with or without borrow in, register only)*
and,or,xor	*(register only)*
asr	*(arithmetic shift right by immediate amount)*
rotl.b	*(rotate left bytes,*
	by true or complement of bottom two bits of register)
shift	*(extract 32-bit field from register pair, immediate shift amount)*
bic	*(bit clear, register only)*
not	*(bitwise logical inverse, register only)*
dstep	*(1-bit restoring divide step)*
mstart	*(the first step of a signed multiply)*
mstep	*(1-bit shift and add step for multiply)*

Memory Transfer Instructions
(all use base plus 17-bit signed offset addressing)

ld/st	*(load/store)*
ldt/stt	*(load/store through, bypassing external cache)*

Branch Instructions
(perform comparison of two registers,
 conditionally branch by 16-bit signed immediate offset.
 Optionally nullify 2 instructions in delay slot if branch not taken)

beq,bge,bhs,blo,blt,bne
                    (branch if equal, greater or equal, higher or same,
                     lower, less than, not equal)

(continued)

Table 37. (continued)

Special Instructions

jpc	*(jump to PC; used in return from exception)*
jpcrs	*(jump to PC and restore state; used in return from exception)*
jspci	*(jump to register + 17-bit signed word offset,*
	save return address to register)
movfrs	*(move from special register)*
movtos	*(move to special register)*
trap	*(trap unconditionally to specified vector)*

Coprocessor Instructions

ldf	*(load floating-point register)*
stf	*(store floating-point register)*
movfrc	*(move from coprocessor to register)*
movtoc	*(move from register to coprocessor)*
aluc	*(coprocessor alu operation)*

(Source: Chow, 1986)

5.2.3 Conclusions

The emphasis on MIPS-X is very much towards the efficient integration of an on-chip instruction cache into the system organization, though the entire architecture has been rethought based on the experience of the MIPS project.

The instruction set is all new, but retains some of the flavour of the original MIPS design. For instance, the use of compare and branch to avoid the need for a condition code register is similar to MIPS. The MIPS-X instruction set and formats also resemble those of the R2000, perhaps not surprisingly.

The MIPS-X pipeline is similar to the R2000 pipeline, though the R2000 employs some additional tricks to allow single delay-slot branches and the use of a physically addressed cache.

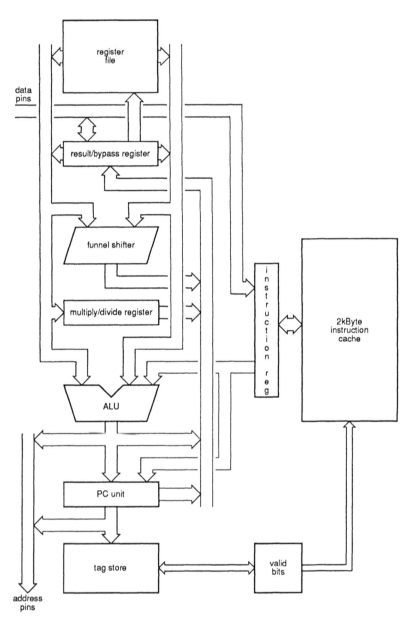

Figure 126: MIPS-X block diagram
(Adapted from Horowitz et al, 1987. ©1987 IEEE.)

5.3 ARM3 - AN ARM CPU WITH CACHE

The 2 micron ARM processor was designed to cycle at 10 MHz worst case, and in the laboratory typical samples operate correctly at 20 MHz. The process design rules used for this design are shrinkable down to 1.5 micron, which should allow an even faster clock rate to be used. Unfortunately, these higher clock rates require a very fast and expensive memory system based on SRAM technology, and are not therefore of great commercial interest.

The usual solution when the CPU can outstrip the memory speed is to introduce a fast cache memory between the CPU and the main memory. This allows the CPU to operate at a higher clock rate whenever the instruction or data item it is fetching is found in the cache, and it only slows to main memory speeds when the item is not in the cache. It was therefore decided to develop a higher performance ARM processor with cache memory which would replace the standard ARM processor for applications which demanded more processing power. The idea was to maintain the same main memory interface so that all the support chips could be used unchanged, but to offer an alternative CPU with higher performance (albeit at a somewhat higher cost). The cache control logic was designed by Alasdair Thomas during 1987; he also developed the cache emulator hardware and software which was used to generate the statistics upon which the cache organizational decisions were based. The circuit design was done by Jamie Urquhart, Harry Oldham and Dave Howard, using VLSI Technology Inc. design tools, and VLSI Technology were responsible for fabrication.

5.3.1 Cache Organization

The major organizational issues for cache memories were raised in chapter 1. In the case of ARM3, the choice between a physically addressed and a virtually addressed cache is determined by the fact that the address translation is performed off-chip in MEMC, so only virtual addresses are available on-chip. The cache therefore has to be virtually addressed.

Since ARM3 uses the same CPU cell as the 2 micron ARM2, it inherits the strongly von Neumann property of the ARM2, whereby it transfers one word of instruction or data every cycle. It is therefore a good match for a single mixed instruction and data cache. Mixed caches tend to balance the proportion of the cache used for data against that used for instructions automatically to suit the current task, whereas a fixed partitioning into separate dedicated instruction and data caches is much less flexible, and performs less well for a given total memory size.

Other decisions were based on a combination of simulations of cache effectiveness and practical considerations. Practicalities limited the maximum size of the cache on the chosen foundry process to 4Kbytes (1K words); simulations showed that it was very desirable to use this maximum size, and even then a good organization was necessary to get a good hit ratio. For instance, a 4Kbyte fully associative cache gave a better performance than a direct mapped cache of 16Kbytes, four times the size.

The cache simulations were performed using a standard ARM system, with a hardware add-on which stored information on every memory reference as it happened. When the information store became full, the ARM was interrupted, and the interrupt routine processed the stored information to generate statistics. Any cache organization could be modelled by the software, and the run-time accumulation of statistics allowed much longer programs to be analysed than would have been practical with off-line analysis of an address trace file.

The various cache organizations were analysed for a range of cache sizes, and with three possible load options. Opcode only (O) caches would only allocate cache entries to items fetched as opcodes, though a subsequent data reference to a location previously fetched as an opcode would use the cached value. Opcode plus data read (O+R) caches would also allocate an entry for a data read item, and opcode plus data read and write (O+R+W) caches would also allocate an entry for written data words (but not bytes). Written data was always copied to main memory; the write-through strategy was adopted for all models to avoid the complexities of cache flushing mechanisms.

Several programs were used for analysing the performance of the various configurations, but one of these seemed to depend more critically on the cache details than the other programs used. The results presented here are all based on this program, which is an interpreted BASIC floating-point benchmark routine. The results are all translated into the ratio of the performance of a 16MHz ARM3 to the performance of a 2 micron ARM, assuming that both processors are attached to identical 8/4MHz MEMC controlled memory systems (with all the bandwidth available to the CPUs). The cache organizations which were analysed initially were:

- *Perfect cache* (figure 127). Every memory read is a hit. This was used as a reference point against which real caches could be compared. Note also the effect of an op-code only cache (O) compared with a data only cache (R). Op-codes are clearly the most important items to cache, but a merged cache (O+R) is the clear winner. Since all reads are found in the cache, the copy on write (W) option is irrelevant here.

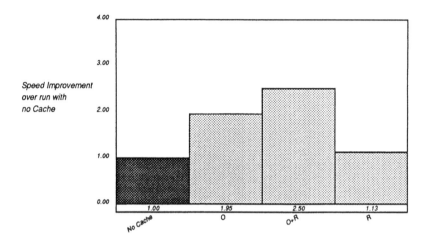

Figure 127: Perfect cache performance

- *Direct mapped cache* (figure 128). The entry a particular memory item is allocated depends only on the low order address bits, and contention arises in small caches between different items with the same low order bits. Even at 16Kbytes, this simple cache organization is well short of the perfect cache performance.

- *Dual-set associative cache* (figure 129). The low order address bits are used to select a pair of possible entries; which one of the pair is used is decided by a least recently used (LRU) algorithm. One bit per entry pair is all that is required to record which of the two was last used, so the LRU implementation can be quite simple. The 16Kbyte cache has near perfect performance, but the maximum realizable size (4Kbytes) gives less good results.

- *Fully associative cache with least recently used replacement algorithm* (figure 130). A memory item may occupy any cache entry; an LRU algorithm is used to determine which entry to re-use next. With this organization the 4Kbyte cache gives the same performance as the perfect cache. A hardware implementation of this cache organization requires an efficient VLSI structure for the LRU algorithm, which is quite complex when usage records are required for several hundred entries.

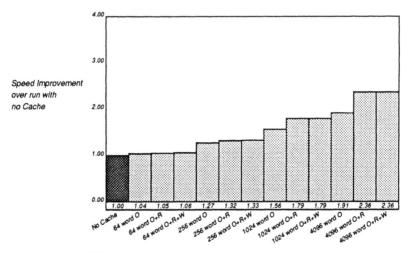

Figure 128: Direct mapped cache performance

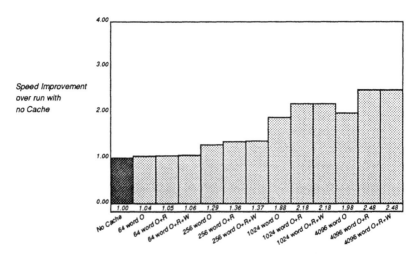

Figure 129: Dual-set associative cache performance

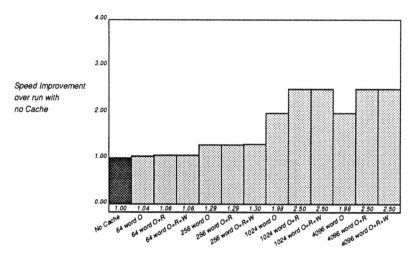

Figure 130: Fully associative LRU cache performance

- *Fully associative cache with random replacement algorithm* (figure 131). Again a memory item may be placed in any cache entry, but now the entries are selected for re-use at random. (In practice either a pseudo-random binary sequence generator is used, or a counter is clocked very rapidly and the pseudo-randomness results from the unknown time at which the value of the counter will be sampled.) This organization works as well as the LRU variant, and is much simpler to implement.

The results show clearly the advantage of a fully associative organization at the practical (4Kbyte) cache sizes. (The performance will be less sensitive to organization at larger sizes, since all organizations approach the perfect figure as size increases.) The random replacement algorithm is much easier to implement than the least recently used algorithm, and performs as well. Also, the LRU cache has an obvious pathological case when a program loop is just larger than the cache; it will always miss. The random cache will perform less well on loops which are just smaller than the cache, but will degrade gracefully as the loop size increases beyond the cache size. Therefore the decision was made to go with a random replacement fully associative cache for ARM3.

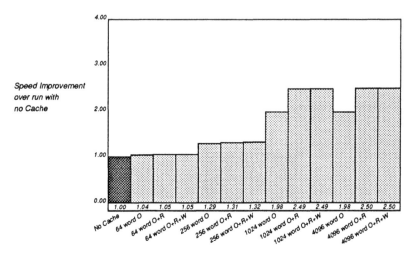

Figure 131: Associative random replacement cache performance

The tag store size is too great if there is one tag for each cache word, so a quad-word line was investigated (figure 132). This reduces the performance of the 4Kbyte cache slightly, but dramatically reduces the chip area required for the tag store.

The O+R cache performs as well as the O+R+W cache, and has better implementation characteristics. If the unit of memory transfer (block) size is the same as the line size, an O+R cache needs only one valid bit for the whole line, and this can be implemented easily as part of the tag CAM. The hit/miss logic uses only the CAM, and gives an early result to the control logic. An O+R+W cache requires a valid bit for each word, which is most logically implemented as part of the data RAM. It therefore requires more silicon area and gives a later hit/miss result. The O+R cache is therefore the best choice for this application. (This is a consequence of the earlier decision to use a *write-through* strategy; a *write-to* O+R+W cache would normally be expected to outperform the O+R cache.)

During the design process it turned out that the power consumption of a 256 entry CAM fast enough to be the tag store of such a cache was too high, so the dependence of the performance on the degree of associativity was investigated further. A 4Kbyte O+R quad-word line (and quad-word block) cache was modelled with associativity varying from full (1 Set/256 Entries)

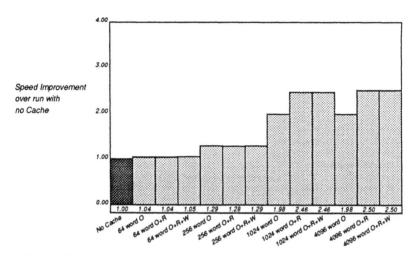

Figure 132: Associative random replacement cache with 4-word line

to none (256 Sets/1 Entry), and the results are shown in figure 133. Note that for these simulations, the cache model had been improved to include more implementation details such as clock re-synchronization times, so all the numbers are somewhat reduced compared with previous figures. These results confirm that going to 64-way associativity has no impact on performance, and allows the tag CAM to be split into four 64 entry blocks, only one of which need be enabled for any access. This reduces the average power consumption to a satisfactory level. It is interesting to note also from this figure that any further reduction in associativity does adversely affect performance.

The final choice of organization is shown in figure 134. The data is kept in a single word-wide block of RAM, and the tags in four 64 entry CAMs. Two low order address bits determine which CAM is active for a given access, and the other CAMs are turned off to conserve power. The cache always reloads a complete line of four words when an op-code or read data miss occurs (which is actually the most efficient use of MEMC controlled DRAM memory, since MEMC uses DRAM page mode within quad-word blocks of memory), so individual valid bits are not necessary for each word (or byte). The hit/miss decision can therefore be made by the CAM alone, which is faster and simpler than it would be if RAM valid bits were factored into the hit/miss decision logic.

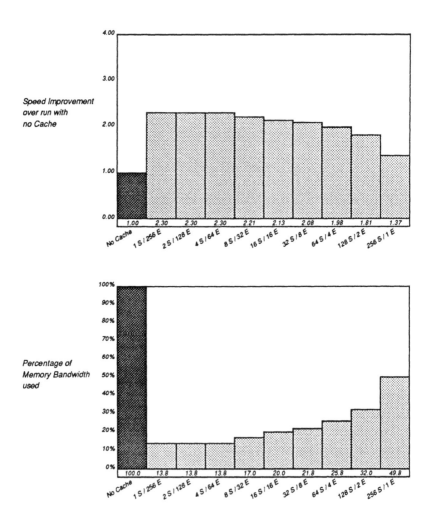

Figure 133: The dependence of cache performance on associativity

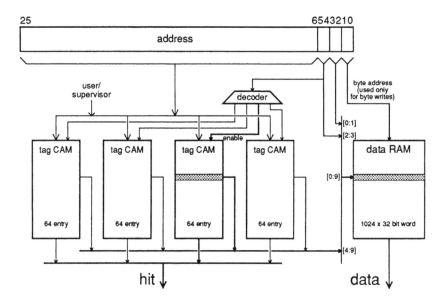

Figure 134: The chosen ARM3 cache organization

The CAM contains the address tag, a valid bit for each quad-word line (all valid bits may be cleared in parallel for cache flushing operations), and two supervisor access bits. One supervisor access bit is used when the supervisor uses the same address translation table as the user (eg when MEMC is used), the other allows separate user and supervisor spaces. In both cases the user is prevented from accessing a cache entry which was fetched while the CPU was in supervisor mode; the access is sent out to the memory translation unit to determine whether it was permissible. If the address space is shared and the access is permissible, the access control is changed and the user subsequently has free access to that entry.

No process identification is maintained in the cache. It is assumed that with a cache this small, a process will use nearly all the space before it is switched out, so an old process will be almost entirely overwritten in the cache. Therefore there is very little to be gained from maintaining process IDs with entries; the scheduler should simply invalidate all the old process entries in the cache before activating the new process.

Each data write is checked against the cache entries, and if it matches one, the cache is updated (including single byte writes). The cache is therefore always consistent with main memory, and self-modifying routines will work exactly as on the uncached ARM. Writes do not affect the cache unless they match existing entries, and they all update main memory with the processor locked to the external memory clock. A write buffer would allow the CPU to continue without waiting for memory, but this would require a completely new translation exception mechanism. The simpler memory synchronous solution has been adopted to enable the existing translation exception hardware and software mechanisms to be used.

An additional benefit of the ARM3 cache is that although the CPU throughput goes up, the external memory bandwidth requirements are greatly reduced (figure 133). A standard (uncached) ARM system is memory bandwidth limited, and the video subsystem (controlled by VIDC) shares the available bandwidth with the CPU. Up to 30 percent of main memory bandwidth is used to support the display, with a corresponding reduction in CPU performance. ARM3 has much lower bandwidth requirements, and is therefore less affected by the display. ARM3 can outperform an uncached ARM by a factor of three in a system where VIDC is using 30 percent of the memory bandwidth. The low bandwidth requirement of ARM3 also makes an ARM3 multiprocessor an attractive possibility.

5.3.2 Cache Control Registers

The cache control registers allow selected areas of the memory map to use the cache in special ways. For instance, it should be disabled for areas of the memory space used for I/O. The 64Mbyte address space of the ARM is divided into 2Mbyte sections, and each section has a control bit in each of three control registers. The control registers can specify:

- *Cacheable*. No cache entries will be generated for accesses to regions which are made uncacheable, which will include I/O areas, and areas of RAM which are double mapped to support hardware scrolled displays.

- *Updateable*. Read accesses are cached, but writes are not copied into the cache for regions of memory where update on write is disabled. This is used for areas where read and write operations are separately decoded; for instance MEMC uses the top of memory for both ROM (on read cycles) and the translation tables (which are write-only). Writes to the translation tables will not be copied to the cache, where they might otherwise corrupt ROM entries.

- *Disruptive.* Any write to a region with disrupt enabled will cause a flush (invalidation of all entries) of the entire cache. This will typically be active over the memory region occupied by the translation tables, where a change to the tables may have invalidated some cache entries, and a complete flush ensures that those entries will be fetched from main memory (and the translation and access control re-checked) before they are next used.

With appropriate initialization, these control options allow ARM3 to be fully code compatible with an uncached ARM; a more sophisticated operating system might benefit from more intelligent cache flushing rather than flushing automatically every time the translation tables are touched. A process change will always require a cache flush, but a simple page swap does not cause cache entries to become invalid, and therefore should not cause flushing even though it requires the translation tables to be modified.

In addition to these control registers, there is also a register containing a master cache enable/disable control bit, a bit to define whether user and supervisor states use shared or separate translation tables, and a bit which turns on a monitor mode for hardware debugging purposes. Writing to another logical register causes the cache to be flushed. A system reset disables the cache, and the control registers should be initialized before it is enabled. A further register contains information identifying the part type, number and version.

The cache control state is modified by the use of coprocessor register transfer instructions addressed to coprocessor number 15, the system control coprocessor. The allocation of the coprocessor register space is shown in figure 135. These on-chip registers are relatively easily interfaced by this means; the interface for off-chip coprocessors is somewhat more complex.

5.3.3 Chip Organization

The major components of ARM3 are shown in block diagram form in figure 136. The functions of the various blocks are described below:

Processor Core (CPU)

The ARM3 processor core is basically an ARM2 CPU with the pads stripped off. The only change to the CPU is the addition of a semaphore instruction (replacing a previously undefined portion of the instruction space). This was felt to be a worthwhile addition, since it required a change only to the instruction decoder PLA, and greatly simplifies the control of shared

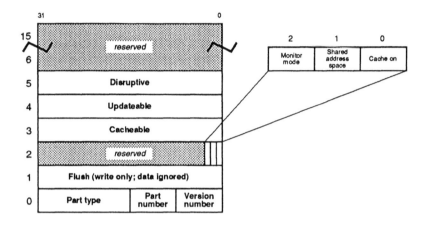

Figure 135: The ARM3 cache control registers

resources in an ARM3 multiprocessor system. The cache on ARM3 reduces the external memory bandwidth requirements of the processor sufficiently to make a shared bus ARM3 multiprocessor a practical proposition.

The instruction set is summarized in figure 137. Key features are:

- All instructions are conditional, to allow the removal of short forward branches from code.

- The data processing operations use a three address format, and allow a general shift of one operand in the same cycle as the ALU operation. Setting of condition codes is optional.

- Multiply uses a two bits per cycle Booth's algorithm to produce the least significant 32 bits of a 32 bit by 32 bit signed or unsigned product.

- The semaphore instruction does an atomic read-then-write of a memory location (and will access main memory even when the addressed location is present in the cache).

- The single data transfer instruction supports base-index and base-offset addressing modes, with optional auto indexing. It loads or stores a single byte or word.

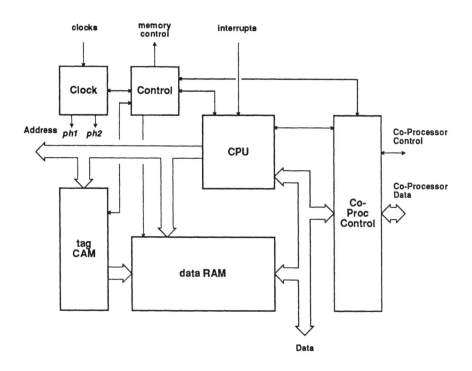

Figure 136: ARM3 block diagram

- The undefined instruction class traps for software emulation of future instruction set extensions.

- Block data transfers allow any subset of the 16 visible registers to be loaded or stored very efficiently. All stacking modes are supported.

- Branches have a 24-bit word offset, and can therefore reach anywhere in the 64Mbyte address space. The program counter may be copied into a link register (R14) for a future subroutine return. Delayed branching is not used; a branch causes a CPU pipeline flush. This approach was adopted to avoid the complexity of restarting delay slot instructions which abort.

- Coprocessor data transfers allow the loading and saving of single and

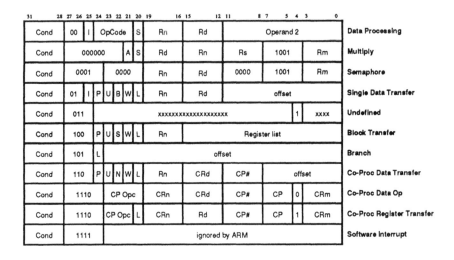

Figure 137: ARM3 instruction summary

multi-word operands between the coprocessor registers and main memory.

- A coprocessor data operation issues an instruction to a coprocessor to perform an internal operation which may overlap with subsequent CPU activity.

- A coprocessor register transfer moves a single word between a coprocessor register and an ARM3 register.

- A software interrupt causes a trap to supervisor code.

The Cache Data Store (Data RAM)

The data store is a 1K by 32-bit static RAM which is byte addressable for writes, otherwise it is word addressed. It is active during the second half of each CPU cycle.

The Cache Tag Store (Tag CAM)

The tag store contains a virtual address for every four words in the data store. It is built as 4 content addressable memories (CAMs), so that when a new virtual address is presented it is compared simultaneously with every entry in one of the CAMs. A matching location will signal a hit, and the CAM will output the address of the location matched. This block is active during the first half of each CPU cycle.

The Cache Control State Machine (Control)

The cache control logic manages the loading of quad-words into the cache, and transforms the asynchronous control signals from the processor core into the same external synchronous interface signals as used by an uncached ARM. In particular, this allows an ARM3 to be connected to a MEMC controlled memory system.

The state diagram is shown in figure 138. The system is forced into the *Reset* state by a system reset, and upon release of the system reset drops into the normal *Fast Cycle* state, where it will stay until an external memory access is required. If the external access is a read of a cacheable area, the *Line Fetch 0..3* sequence is entered; otherwise an *External Access* is performed.

Since read data is valid only at the end of a memory access cycle, an additional *Write-Back* cycle is required to copy the last word of the quad-word block into the cache. This may either be *Slow* if another line fetch is to follow, *Slow External* if an uncacheable memory access is to follow, or *Fast* if the cache is to be used or if the nature of the following cycle is unknown at the decision time.

The *Line Update* state is entered when the memory access is only to check whether an existing cache entry (which had been fetched in supervisor state) is user accessible, and the *Dummy Fetch* state is used to prevent such an operation (or a line fetch which aborts) from corrupting the cache.

The Coprocessor Control Logic (CoProc Control)

The ARM coprocessor interface is optimized towards high performance, and uses a bus watching protocol whereby the ARM and the coprocessor take a copy of each instruction as it is fetched from main memory. While the ARM decodes the instruction and determines whether it is to be executed (remember that all ARM instructions are conditionally executed), the coprocessor decodes it to determine whether it can commit immediately. The handshake to commit execution can take just the one cycle required by ARM to step its instruction pipeline, and this is the only overhead for executing the instruction. If the coprocessor is busy, or requires time to prepare data items requested by the instruction, the handshake puts the ARM into an interruptable busy-wait state until the coprocessor is ready.

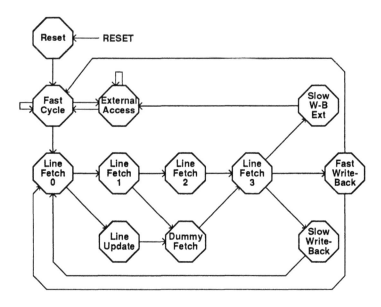

Figure 138: The cache control state diagram

The problem for a coprocessor connected to a cached CPU is that the CPU no longer collects all instructions from main memory; most of them come from an invisible on-chip cache. The coprocessor bus watch mechanism therefore requires rethinking. The obvious solution is to attach the coprocessor to the bus between the CPU and the cache, down which all the instructions flow. The problem is that this requires additional pins to bring this bus off the chip, and the CPU clock rate is likely to be so high that sending every instruction fetch off chip at the CPU clock rate will require a lot of power. Additionally, the delay in getting the instruction off the chip and into the coprocessor will either compromise the on-chip timing or the coprocessor timing skew.

The solution adopted on ARM3 accepts the cost of the additional bus, but puts a one clock pipeline delay into the route off-chip. This additional cycle is long enough for the coprocessor instructions to be identified, and the bus drivers can be disabled for all others, saving the unnecessary power dissipation. The coprocessor therefore runs one cycle behind the ARM, and the minimum handshake time increases to two (fast) clock ticks. Data transfers from main memory to the coprocessor spend one tick on ARM3 to match this delay, and this keeps memory timing very safe. Data transfers from the coprocessor to main memory must be timed to allow for the delay, but again pipelining is used to ensure safe timing.

The ARM3 interface uses extra cycles to ensure consistent memory timing, but the increased CPU clock rate means that the actual handshake time is reduced. As a result of this implementation, the coprocessor is also able to exploit the increased memory bandwidth available from the cache.

Clock Selection and Synchronization (Clock)

ARM3 has two asynchronous clock signals; one is locked to external memory cycles, the second determines the rate of execution when all accesses are in the cache.

ARM data transfers are synchronous to the CPU clocks. On ARM3 a standard CPU core is used, but now it must accept transfers from either the on-chip cache or the off-chip main memory. Therefore the CPU is made synchronous to the data source for each cycle, and the CPU clocks switch between the external memory clock and the fast clock (for cache accesses). Furthermore, to maximize flexibility in the choice of the fast clock rate, it was decided to allow the two clocks to be completely unrelated. Every switch between the two clock frequencies requires resynchronization, with the unavoidable possibility of synchronization failure and metastability in the clock selection circuitry.

The pressure on the synchronizer design is high because the resynchronization time is mostly wasted time, and should therefore be minimized, but the safety of a synchronizer increases exponentially with the synchronization time. Finding the optimal trade-off point for safe operation with minimal delay requires very careful engineering design.

5.3.4 Cache Implementation

The design of ARM3 was targetted at a standard CMOS foundry logic process, which has no special features (buried contacts, high resist polysilicon, etc.) to improve RAM density. Double level metal was available, and this greatly simplifies power distribution and the control of clock skews across the chip. All design work had to allow for the significant process tolerances which apply to all such general-purpose foundry processes.

The Tag CAM

The tag CAM is built from a 9 transistor CAM cell (see figure 139) with dynamic precharging, and self-timing circuits are used to give an early reliable hit/miss output and a glitch free match address. The CAM is split into four separately enabled blocks of 64 entries to reduce power dissipation. In total, the tag CAM uses 69,500 transistors and 17 square mm of silicon area on a 1.6 micron process.

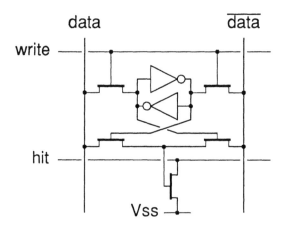

Figure 139: The tag CAM cell

The Data RAM

The data RAM uses a conventional 6 transistor SRAM cell, and a polysilicon word line with metal strapping to improve access time. It is implemented as two blocks of 16kbits (64 by 128 bits), and uses analogue differential sense amps. No address transition detection is required because of the synchronous nature of all the relevant on-chip signals, but a self-timed circuit is used to power down the sense amps when the outputs have been latched. The block uses 203,400 transistors in 25 square mm on a 1.6 micron process.

5.3.5 Testability

Chips with 300,000 transistors can be very hard to test in production, and testing can become a major cost item. Cached CPUs can also present difficulties for software debugging, as a great deal can happen inside the chip with no way of knowing what is happening from the outside. ARM3 has a number of features to ease testing and debugging of the chip, both in isolation and in circuit:

- Test modes on the chip allow the CPU, RAM and CAM blocks to be tested independently, with direct access to RAM address and data lines, etc.

- In normal operation, when the processor is executing from cache, the address lines are static to conserve power, and the coprocessor bus is only driven when coprocessor instructions are being fetched (or data is being transferred to the coprocessor). In the software selectable monitor mode, the internal addresses are always driven onto the external address bus to allow program tracing by a logic analyser, and all data traffic is driven onto the coprocessor data bus. Also, the processor runs off the memory clock for all cycles (including cache cycles). The continuous driving of the outputs will increase the power consumption of the chip, though the lower clock rate will alleviate this effect, and this should be acceptable during debugging. This feature will greatly simplify the tracking down of obscure program bugs, by making all processor activity externally visible.

5.3.6 An ARM3 System

Figure 140 shows the system block diagram for an ARM3 system built with the standard ARM support chips. The only difference from the standard ARM system is the substitution of the new processor, and the connection of the coprocessor onto the dedicated coprocessor bus instead of onto the main bus. (ARM3 also generates a new clock for the coprocessor interface, which operates at a higher rate than the corresponding clock in the uncached system.) This system delivers around 8-10 MIPS, which is 2 to 3 times the performance of the system with the standard ARM, for a modest cost increase over the standard ARM system.

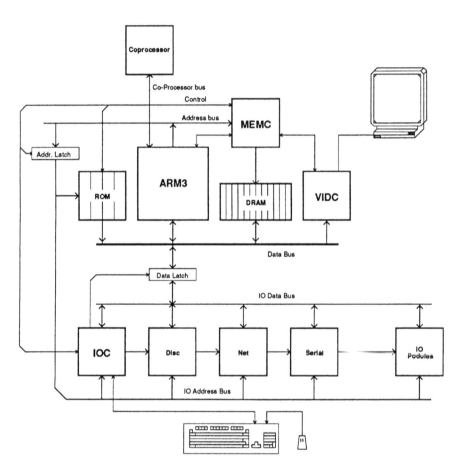

Figure 140: ARM3 system block diagram

5.3.7 Concluding Remarks

Even a relatively simple cache implementation requires careful consideration of speed, cost, performance, power, implementation and organizational issues. The most important considerations on ARM3 were:

- *Organization*. The importance of high associativity was recognized, with random allocation and quad-word line length being compromises which had little performance impact but great cost benefit.

- *Power*. All outputs are turned off as much as possible, and the tag CAM is divided into four blocks, only one of which is active at any time. Auto power-down of the RAM sense-amps also helps.

- *Testing*. The chip can be divided into separately testable blocks which are already understood from a testing point of view (RAM, CAM, ARM core). The optional monitor mode gives visibility of internal operations at the cost of added power dissipation (and slower running).

5.4 ANALYSING ARCHITECTURAL DECISIONS

The designer of a computer architecture and organization is faced with a bewildering array of decisions, and very little solid ground upon which to base them. If there were a clear way of establishing that the best number of registers for a RISC processor is 32, then all designers would quickly choose 32 and worry about something else. A quick look at chapter 3 will show that there is no such uniformity of decision on the number of registers, or on anything else for that matter.

The main way of making architectural choices is on the basis of precedent. If it worked before in a successful machine, it can't be completely wrong. Just occasionally, a radically new idea (like RISC) is accepted for the basis of a new design, but such occurrences are rare. It is very difficult to evaluate all the implications of a design feature, and it is very costly to get it wrong.

Take, for instance, the Berkeley windowed register file. The arguments for it are clear and quantifiable (much reduced data traffic, faster procedure entry and return), as are those against (increased silicon area, slower register access, slower context switches when all the registers must be saved). In the case of the Acorn RISC Machine the decision was easy; this feature could be ruled out on cost grounds alone because of the increased silicon area. But for higher cost points the position is less clear. With a fixed silicon budget, the register file must be traded off against something else. A larger register file is a good thing if it comes for free, but what if it comes in place of an instruction cache, for example?

Recently some tools have become available which give system designers the capability of experimenting with architectural features. Flynn, Mitchell and Mulder (1987) have used such a tool, called a *computer architect's workbench* (Mitchell, 1986), to evaluate the impact of certain instruction set, register file and cache decisions. They come to the following conclusions:

(1)   A simple uncached processor with up to 16 registers, a 32-bit instruction format and a load-store architecture is a reasonable design point for a low-cost processor. The simplicity is reflected in a 50% to 100% higher instruction bandwidth than would be required by a processor with an instruction set which used a more complex encoding to minimize the bandwidth.

(2)   If the simple processor is extended to reduce the instruction traffic, a denser instruction set should be adopted before an instruction cache is added. For a given performance level, the instruction cache for the

denser instruction set can be 50% smaller than that for the simple instruction set.

(3) For minimizing data traffic, a 16 register file with a small data cache is better than the equivalent silicon area occupied by a large windowed register set. (The difference is actually quite small, so both solutions are likely to persist!)

The report also comments on the performance impact of the number of registers available to the compiler. The conclusion here is that 16 registers are sufficient for most purposes, except when a compiler performs inter-procedural optimization, when it can make good use of many more than 16. However inter-procedural optimization is not commonly used because it conflicts with the recompilation of separate program modules; normally only global optimization within a procedure is performed.

The overriding conclusion is to warn against the trend to simple instruction sets at the cost of code density. The code density determines the memory bandwidth requirements and the effectiveness of the instruction cache, and ultimately these are likely to determine the processor performance, not the ease of decoding. The RISC approach should not be taken too far!

5.5 THE LIMITS TO PERFORMANCE

The introduction of RISC architectures has been accompanied by a very rapid rate of increase in available VLSI CPU power. Every few months a new processor is introduced to the market place with higher performance than its predecessors. Will this continue indefinitely? If not, where are the limiting factors which will slow this process?

If system performance is the issue, then there are three configurations to consider:

(1)   *Uniprocessor systems.* The most common organization for a computer uses a single CPU for general-purpose computing. It may contain other embedded processors for graphics or I/O. This configuration can have a tightly coupled main memory, and minimal logic is required for cache consistency. The computing power of the system is determined by the single main processor.

(2)   *Shared memory multiprocessors.* Here several CPUs cooperate to perform the user tasks. They are all attached to a common bus, and operate from the same main memory. Each processor has a cache to reduce its bus bandwidth requirements; without this the bus could not support all the CPUs. A hardware cache consistency mechanism is usually employed to allow system software to operate under the illusion of a single shared memory space. The memory response time to any individual processor is normally worse than for the uniprocessor, because time is required for bus arbitration, and there will sometimes be contention for the bus. The limit to system performance is usually determined by the number of CPUs which the bus can sensibly support before becoming saturated, and the performance of each CPU.

(3)   *Communication based multiprocessors.* A large number of CPUs may be configured as a single computing unit if each has its own memory, and a communication network allows the various processes to exchange messages. A lot of research has gone into these configurations, and the Inmos transputer is a VLSI implementation of a node for such a system. The configuration can be extended indefinitely, but the performance for a particular application will be determined by how well it can be broken down into parallel elements, and how well the communication system meets the needs of the application in question. Such configurations are still used mainly for special-purpose computations, and will not be discussed further here.

System types 1 and 2 share many of the same properties when it comes to optimizing the design of their CPUs. The components which are required in each processing node of a typical multiprocessor are shown in figure 141. A uniprocessor node is similar, except that the bus snooping logic may be unnecessary and the bus interface is less complex. There may be one cache, or separate ones for instructions and data. The TLB may be folded into the main cache as described earlier for the Berkeley SPUR project. The FPU may be absent for applications which do not require it.

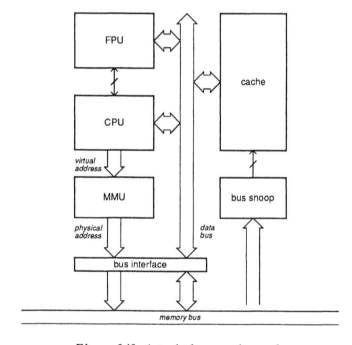

Figure 141: A typical processing node

The processing node may be partitioned in many ways. The designs described in chapter 3 include chips with the FPU on the CPU chip and the cache and MMU split into instruction and data sections on a chip each, or chips with a small instruction cache included. Some include multiprocessor bus interfaces, others allow this to be added externally. The trend in VLSI is to put more on a single chip, so how will this node be partitioned in the future?

It is possible to extrapolate from present practice into the near future to see what sorts of VLSI implementations might become available. We start with a number of assumptions about the system, and consider a CMOS VLSI processor with an on-chip cache:

(1)   Assume that CMOS has a gate delay inversely proportional to the feature size. This is approximately true for long metal tracks, where fringe capacitance effects are dominant at small geometries. On a 2 micron process, a 32-bit ALU can be built with a 20 to 30 nsec delay, so a clock rate of 25 MHz is possible. Therefore, for a future chip based on a process with a gate length of f microns:

clock rate = 50/f MHz ; clock period = 20 f nsec.

(2)   Assume that the on-chip cache size is inversely proportional to the square of the feature size, and the average miss ratio halves for every quadrupling of the cache size. (This is very approximately true for cache sizes in the range 10 to 100Kbytes.) If a cache on a 2 micron process is around 4Kbytes, and has a miss ratio of 5%, for our future chip:

miss ratio = .025 f.

(3)   Assume that the main memory system returns a cache line (four words) of data in 400 nsec. This is consistent with a main memory based on DRAM which is 32 bits wide, and each miss reloads a cache line using a burst access mode. The average CPU cycle time is the clock period plus the product of the miss ratio and the miss cost:

average cycle time = (20 + 400 x .025) f nsec.

(4)   A good RISC processor might approach one instruction per cycle:

uniprocessor performance = 33/f MIPS.

(5)   The uniprocessor requires the memory bus whenever it misses the cache:

bus use per microsecond = 400 x 50/f x .025 f nsec = 500 nsec.

Therefore the bus can only support 2 processors before it saturates. On the basis of the above assumptions, the bus requirement is independent of the processor feature size. As the geometries shrink, the faster CPU clock just balances the reduced miss ratio which results from the larger cache memory.

On a near-future 0.8 micron process the uniprocessor delivers 40 MIPS. It operates at 60 MHz, and has a 25Kbyte cache. The dual processor gives up to 80 MIPS. On this basis the multiprocessor is not very attractive, and would benefit from additional off-chip cache memory to reduce its bus bandwidth requirement. If the bus were made faster, probably by making it 64 bits wide, then the uniprocessor would go a little faster (due to the reduced miss cost) but the multiprocessor would benefit more by allowing three or four processors to be used at once.

On such a chip the CPU part would be a very small feature. Even on ARM3, the processor uses only a quarter of the chip area. The memory will use most area. It therefore makes sense to optimize the process for the memory rather than for the CPU. A special memory process would allow a larger cache to fit in the same area, affecting the above uniprocessor/multiprocessor balance considerably. If the cache were made from the highest density memory available, which is dynamic RAM, then a very large cache would be possible. Unfortunately DRAM is rather slow, even for accesses on the same chip, but a small static RAM cache could be used as well to compensate for this. Highly integrated processors will increasingly look like memory chips with added intelligence, a fact that Inmos have been taking account of for some time. The transputer is based on a process optimized for the density of the on-chip static RAM.

The above 'back of the envelope' calculation is sensitive to a number of assumptions, most notable of which is the cache hit ratio. Cache organization is very important - the ARM3 4Kbyte associative cache gave the same hit ratio as a direct mapped cache of four times the size. The cache hit ratio is also very sensitive to the program being run. Perhaps compilers will develop towards better use of caches. As a simple example, the linked-list structure which is widely used by programmers could have been designed to break cache algorithms. There may be equivalent ways of managing the list which are less efficient on uncached machines, but which display better reference locality and therefore work better on machines with caches.

5.5.1 Power Dissipation

The power consumed by a chip is often a performance limiting factor. The logic components have approximately constant strength transistors as the process shrinks; they get faster as a result of the reduced capacitive loading. Therefore the power consumed by the integer unit or the FPU is independent of the feature size. The cache memory power is determined by the number and speed of the sense amplifiers, and the switching of long signal tracks.

Although the memory array uses many more transistors as it gets larger, roughly the same number are active in any access. The power therefore depends mainly on the clock rate, so it will go up inversely in proportion to the feature size. Faster processors will continue to get hotter mainly due to their faster cache access cycles.

If the aim is to minimize power dissipation, then instead of using the maximum clock rate and cache size allowed by the chip size, these factors could be balanced differently. A maximum speed 2 micron chip delivers 15 MIPS, probably at around 2 watts. If simply shrunk onto a 0.8 micron process without changing the cache size or clock frequency, the power should reduce to 800 milliwatts. A larger cache will increase the CPU dissipation, but might save power overall by reducing the number of main memory accesses.

5.5.2 The Floating-Point Unit

As clock rates increase, communication with a tightly coupled off-chip floating-point unit gets increasingly difficult and power consuming. Floating-point operations are often communication rather than computation limited, and it can be beneficial to bring the FPU onto the CPU chip even if to do so means reducing the floating-point computational power. Beyond 50 MHz clock rates, the FPU will either be on chip, or use special inter-chip connection technology.

5.5.3 The Future of RISC

From the above, it may be concluded that two extreme configurations of RISC processor node might evolve in the next few years, on 0.8 micron CMOS or similar processes:

(1)   The all-out performance processor for multiprocessor machines. This will use a multi-chip organization, deliver 50 to 100 MIPS per node, and dissipate considerable power. The chips will be connected on a special substrate to make interchip communication possible at 100 MHz clock rates, and may use ECL or Gallium-Arsenide technologies for some components. Nodes will have on-chip caches to reduce pin bandwidth, and off-chip caches to reduce bus bandwidth. The machines will use up to 16 processors on a common bus, and deliver around 1000 MIPS in all.

(2)   Highly integrated uniprocessors will be developed for more cost-sensitive applications. These will perform relatively poorly as

multiprocessor nodes, since their on-chip caches are too small, so they use too much bus bandwidth. Their uniprocessor performance will, however, approach that of the multiprocessor node. They will have 32Kbyte on-chip caches, and on-chip floating-point hardware where needed. Some will be aimed at the 50 MIPS low-end workstation market, and others at embedded controllers. Low power versions will appear in 20 to 30 MIPS portable computers, and as ASIC library cells for custom controllers.

All the above machines will offer considerable processing power at low cost by present standards. The system designers will spend most of their time attempting to match the I/O performance to the CPU power.

Once these configurations have stabilized, performance will increase only as a result of improvements in semiconductor processing. The RISC approach has generated a phase of development where advances have come much faster than would have resulted from simply tracking improvements in processing, since over this period the number of cycles per instruction has reduced from the CISC norm of around five to the RISC goal of just over one. The basic clock rate has remained linked to processing, being based on the cycle time of a 32-bit ALU. The next advance in CPU performance awaits the next conceptual breakthrough. How do we get past one instruction per cycle? Perhaps very long instruction word (VLIW) machines will have their day. Perhaps the long-sought general solution to fine-grain concurrency will be found, and the transputer will come into pre-eminence.

In the meantime, there are several hard problems to do with I/O bus architectures which need good solutions, if the RISC CPUs which are already here are to be kept busy doing useful work.

References

Chow, P., (1986). MIPS-X Instruction Set and Programmers Manual, Stanford University Technical Note No.CSL-86-289.

Chow, P. and Horowitz, M., (1987). "Architectural Tradeoffs in the Design of MIPS-X," Proceedings of the 14th International Symposium on Computer Architecture, ACM SIGARCH, Pittsburg, PA, pp. 300-308.

Flynn, M. J., Mitchell, C. L. and Mulder, J. M., (1987). And Now a Case for More Complex Instruction Sets, IEEE Computer, 20, no. 9, pp. 71-83.

Hansen, P. M. and Kogn, S. I., (1986). SPUR Coprocessor Interface Description, Berkeley University Report No. UCB/CSD 87/308.

Hill, M. D., Eggers, S. J., Larus, J. R., Taylor, G. S., Adams, G., Bose, B. K., Gibson, G. A., Hansen, P. M., Keller, J., Kong, S. I., Lee, C. G., Lee, D., Pendleton, J. M., Ritchie, S. A., Wood, D. A., Zorn, B. G., Hilfinger, P. N., Hodges, D., Katz, R. H., Ousterhout, J. and Patterson, D. A., (1986). Design Decisions in SPUR: a VLSI Multiprocessor, IEEE Computer, 19, no. 11, pp. 8-22.

Horowitz, M., Chow, P., Stark, D., Simoni, R. T., Salz, A., Przybylski, S., Hennessy, J., Gulak, G., Agarwal, A. and Acken, J. M., (1987). MIPS-X: A 20-MIPS Peak, 32-bit Microprocessor with On-Chip Cache, IEEE Journal of Solid-State Circuits, sc-22, no. 5, pp. 790-799.

Katz, R. H., Eggers, S. J., Wood, D. A., Perkins, C. L. and Sheldon, R. G., (1985). "Implementing a Cache Consistency Protocol," Proceedings of the 12th International Symposium on Computer Architecture, ACM SIGARCH, Boston MA.

Lee, D., (1986). Data Path Design Considerations for a High Performance VLSI Multiprocessor, Berkeley University Report No. UCB/CSD 87/318.

Mitchell, C. L., (1986). Processor Architecture and Cache Performance, Stanford University Technical Report No.CSL-TR-86-296.

Ritchie, S. A., (1985). TLB for free: In-Cache Translation for a Multiprocessor Workstation, Berkeley University Report No. UCB/CSD 85/233.

Acronyms

ALU   **Arithmetic and Logic Unit.**
Where operands are added, ANDed, etc.

CAM   **Content Addressable Memory.**
A memory where an entry is identified by its contents.

CISC   **Complex Instruction Set Computer.**
A computer with many complex addressing modes and
operations defined in its instruction set.

CMOS   **Complementary Metal Oxide Semiconductor.**
A logic process which uses both p-type and n-type FETS.

CPU   **Central Processing Unit.**
The part of a computer which contains the instruction decoder,
the registers, the ALU etc., but not the memory or I/O systems.

DMA   **Direct Memory Access.**
A mechanism which allows data to transfer between
memory and the I/O system without involving the CPU.

DRAM   **Dynamic Random Access Memory.**
Read/write memory where information is held as stored charge.

ECL   **Emitter Coupled Logic.**
A fast bipolar logic process.

FET   **Field Effect Transistor.**
A switching device based on electric field effects.

FPA   **Floating-Point Accelerator.**
Logic which may be added to a CPU to improve the
performance of floating-point operations.

FPU   **Floating-Point Unit.**
Similar to an FPA, but often with its own registers
for floating-point operands.

I/O   **Input/Output.**
The movement of programs and data between main memory
and peripheral devices (disks, printers, etc.).

LAN   **Local Area Network.**
A communication system which connects computers together.

LRU   **Least Recently Used.**
The item in a cache which has not been accessed
for the longest time.

MFLOPS   **Millions of Floating-point Operations per second.**
A measure of a computer's floating-point performance.

MIPS   **Millions of Instructions Per Second.**
A (crude!) measure of a computer's integer performance.
Also a name associated with a Stanford research project,
and the name of a commercial organization.

MMU   **Memory Management Unit.**
Where virtual addresses are translated to physical addresses;
often contains a *TLB*.

NMOS   **N-type Metal Oxide Semiconductor.**
A logic process which uses only n-type FETs.

PC   **Program Counter.**
A CPU register which contains the address of the current
instruction.

PLA   **Programmable Logic Array.**
A regular AND-OR array which can be constructed to
perform a wide range of combinatorial logic functions.

PMOS   **P-type Metal Oxide Semiconductor.**
A logic process which uses only p-type FETs.

PSR   **Program Status Register.**
A CPU register which contains assorted flags such as
mode bits, interrupt enables, condition codes, etc.

PTE   **Page Table Entry.**
An element of the virtual to physical address translation
table, containing the mapping for one page.

RAM **Random Access Memory.**
Read/write memory where an entry is identified by its address.

RISC **Reduced Instruction Set Computer.**
A computer with an instruction set which has been
optimized for efficient hardware implementation.

ROM **Read Only Memory.**
Memory with a fixed content.

SRAM **Static Random Access Memory.**
Read/write memory based on bistable storage elements.

TLB **Translation Look-aside Buffer.**
A cache of recently used *PTEs*.

TTL **Transistor-Transistor Logic.**
A bipolar logic process.

VLSI **Very Large Scale Integration.**
Integrated circuits containing tens of thousands of devices.

VRAM **Video Random Access Memory.**
DRAM devices with an additional high bandwidth serial
access mechanism, optimized for video display generation.

Index

9 780824 781514